Cowles Foundation
for Research in Economics
at Yale University

Monograph 26

Cowles Foundation

FOR RESEARCH IN ECONOMICS AT YALE UNIVERSITY

The Cowles Foundation for Research in Economics at Yale University, established as an activity of the Department of Economics in 1955, has as its purpose the conduct and encouragement of research in economics and related social sciences with particular emphasis on the development and application of logical, mathematical, and statistical methods of analysis. The professional research staff have, as a rule, a departmental appointment and some teaching responsibility.

The Cowles Foundation continues the work of the Cowles Commission for Research in Economics founded in 1932 by Alfred Cowles at Colorado Springs, Colorado. The Commission moved to Chicago in 1939 and was affiliated with the University of Chicago until 1955. In 1955 the professional research staff of the Commission accepted appointments at Yale and, along with other members of the Yale Department of Economics, formed the research staff of the newly established Cowles Foundation.

A list of Cowles Foundation Monographs appears at the end of this volume.

The Efficient Use
of Energy Resources

WILLIAM D. NORDHAUS

NEW HAVEN AND LONDON YALE UNIVERSITY PRESS 1979

Designed by Thos. Whitridge
and set in Monophoto Bembo type
by Asco Trade Typesetting Ltd., Hong Kong.
Printed in the United States of America by
the Vail-Ballou Press, Binghamton, New York.

Published in Great Britain, Europe, Africa, and Asia
(except Japan) by Yale University Press, Ltd., London.
Distributed in Australia and New Zealand by Book &
Film Services, Artarmon, N.S.W., Australia; and in Japan
by Harper & Row, Publishers, Tokyo Office.

Library of Congress Cataloging in Publication Data

Nordhaus, William D
   The efficient use of energy resources.

   (Monograph–Cowles Foundation for Research
in Economics at Yale University; 26)
   Bibliography: p.
   Includes index.
   1. Energy conservation—Economic aspects.
I. Title.   II. Series: Cowles Foundation for
Research in Economics at Yale University.
Monograph; 26.
TJ163.3.N67   333.7   79-64225
ISBN 0-300-02284-0

To

JEFF, MONICA, and BECKA

CONTENTS

FIGURES

TABLES

ACKNOWLEDGMENTS

Many people have contributed to this study. The use of mathematical programming as a way of performing quantitative analysis has its basis in published works and informal discussions. Tjalling Koopmans of Yale was a pioneer in this field and helped me at several points. Alan Manne of Stanford University and Kenneth Hoffman of Brookhaven National Laboratories wrote the seminal work in the energy area. I benefited from the comments of William Brainard, William Hogan, Herbert Scarf, James Sweeney, and James Tobin.

In addition, a large number of researchers helped with computer programming and other chores. I particularly would like to thank Truus Koopmans, Ludo van der Heyden, Paul Krugman, Robert Litan, and Richard Peck. Nonincriminating institutional support was provided by the Cowles Foundation, the National Science Foundation, and the International Institute for Applied Systems Analysis.

To those about to embark, I can only say I hope you enjoy the voyage as much as I did.

INTRODUCTION

Another book on energy requires apology more than introduction. My basic apology is twofold: questions concerning energy and exhaustible resources are of fundamental importance to our economy and society, and many of them remain unanswered.

The Basic Themes

The fundamental economic question about energy resources is: How fast should we consume our low-cost resources? (Of course, in a mixed market economy such as the United States there are no simple and direct levers to control the rate of consumption of our resources, and we must use policy instruments like interest rates, taxes, price controls or subsidies, and environmental constraints.) From a technological point of view, we face three different kinds of resources. First, there are inexpensive but limited oil and gas resources—ideal from an economic and environmental point of view. Second, there are abundant, but less attractive, resources that may be used when the low-cost resources run thin—coal, high-cost oil and gas, high-grade uranium in the current generation of nuclear power reactors. This second group generally suffers from the shortcoming that it is expensive and often environmentally risky or dirty.

A third group is the superabundant resources that would provide virtually limitless energy for centuries to come—these are fusion, fission, solar, and unknown. These share two features—they are unproven for large-scale use, and they are relatively expensive. Beyond that, some are thought to be clean (solar energy), while some are thought to be environmentally very risky (some nuclear breeder reactors); some are virtually proven (the liquid metal fast breeder reactor), while others have not passed the test of technical feasibility (fusion); some are soft, some hard, and so forth.

This book asks a number of questions about the time pattern of use of these various resources:

1. What does economic theory have to contribute to the question of efficient use of energy resources? Quite a bit, as is shown in chapter 1. Efficient use of energy resources entails using cheap before expensive resources. A deeper set of results concerns the efficient pricing of resources. Each resource will, in a competitive market, have a "royalty" attached to it. The royalty will be zero for resources that are not scarce, positive for those that are. For all resources, the royalty will be rising at the market interest rate.

By working backward from exhaustion, we can determine what an efficient price for oil or other resources would be. The basic result can easily be seen where there are no extraction costs—roughly accurate for Mideast oil today. In this case, at the point when substitution of the next resource (higher-cost oil, coal, whatever) occurs, the price of Mideast oil and its substitute must be equal. For concreteness, call that year 2020. In 2020, then, the royalty on Mideast oil must equal the cost of the substitute. Since, in an efficient market, the royalty must rise at the interest rate, the royalty today must be the discounted value of the royalty in 2020. If the discount rate is 6 percent, and the substitute costs $20.00, then in 1975 the royalty on Mideast oil, *and its efficiency price*, must equal $20.00/(1.06)^{45} = $1.45.

2. How can we apply these simple economic theories to the real world? Clearly, casual examples like those presented in the last paragraph are insufficient to indicate what an efficient price would be in the real world. Chapters 2 through 4 describe the construction and components of a model designed to determine the efficient path for using energy resources.

The model has two different components. The first, discussed in chapter 2, describes the "demand" side of the energy market. It reports the results of a detailed econometric model of energy demand and then shows how these results can be used in the energy model. The second component is the technology, described in chapter 3, which presents estimates of the extent of energy resources, as well as the costs of extraction and conversion. Alternative models of cost of extraction are briefly described. Finally, chapter 4 lists the detailed equations used in the linear programming energy model.

One major spillover from the model construction is the estimates of energy demand functions in chapter 2. These rely on a combination of techniques for estimating the price-responsiveness of energy demanded in the United States and Europe. The basic result is that energy demand is shown to be moderately elastic with respect to price, with elasticities in the range of −0.5 to −1.0 depending on the sector, country, and specification.

3. The most important investigation, in my mind, relates to the estimate of the efficiency price of oil, given in chapter 5. Relying on the model, and the (clearly unrealistic) assumption that the energy market is competitive, we estimate that the efficient price of oil (for 1975 in 1975 prices) is $3.00 per barrel. This compares with a price of approximately $11.00 per barrel in 1978 (again in 1975 prices). The reason why the calculated efficiency price is so surprisingly low is basically the reason given in question 1 above—the cost of the next substitute resource is relatively modest, and the time at which substitution occurs is distant. Extensive sensitivity analysis in chapter 5 gives a range of $2.03 to $3.71 per barrel—still well below the present market price.

4. Given the enormous discrepancy between actual and calculated efficiency price, does this suggest that the Organization of Petroleum Exporting Countries (OPEC) is responsible for the difference? Chapter 1 investigates briefly the theory of monopoly in resource markets. Under limited but plausible assumptions it is shown that the monopoly price will be set at approximately the substitute price. In the example in (1)

above, then, if a monopolist had control of the oil market, he would set the price at slightly below the substitute (say, $19.00), rather than at the competitive price of $1.45.

Is OPEC responsible for the current oil price? It is a tempting hypothesis to attribute the rise of the world oil price from 1972 to 1974 to the effective monopolization of the world oil market by OPEC. This temptation is reinforced by the result in chapter 5 that the market price in the late 1960s and early 1970s was virtually equal to the calculated efficiency price. In chapter 6, we look more carefully at the empirical support for this hypothesis, both in the current study and in other studies. Most studies make a motivational hypothesis that OPEC is interested in maximizing its discounted profits (the "wealth maximizing monopolist"). The basic result of this and other economic studies indicates that the wealth-maximizing price for OPEC oil today lies at the bottom end of the $10.00 to $20.00 per barrel range (in 1975 prices). These studies confirm that the price rise of 1973–74 can be traced basically to the virtual monopolization of the international oil market. Although a large degree of short-run monopoly power exists, and given the undeniable presence of irrational elements in oil pricing, the evidence strongly suggests that any further substantial rise in the oil price would take prices well above the long-run monopoly price.

5. How likely is it that the future will see a repetition of the dramatic price rises of 1973–74? Chapter 6 describes the trajectory of wealth-maximizing oil prices for this model as well as for other studies. Although there are enormous uncertainties, these studies indicate that on economic grounds the 1973–74 increase attained the long-run monopoly price. Taking the median of the studies presented there, the real price of oil that would maximize the wealth of OPEC would rise about 2 percent per annum over the period 1975 to 2005.

6. What is the chance that global environmental effects will appear as a result of unrestrained market forces? There is widespread evidence that the combustion of fossil fuels—leading to buildup of atmospheric carbon dioxide—will be the first man-made environmental problem of global signi-

ficance. Chapter 8 outlines the essence of the problem and
investigates possible control strategies. The conclusion is that
we are probably heading for major climatic changes over the
next 200 years if market forces are unchecked. Global temperatures would rise in the order of 5 degrees centigrade, with
much more dramatic increases in the polar regions. Further
consequences—on agriculture or the level of the oceans—
would clearly follow.

Chapter 8 describes several control strategies, of which
the most efficient one is a "carbon tax" on the combustion
of fossil fuels. It is shown that (depending on the desired level
of control) these can reduce the seriousness of the effects with
modest but manageable economic costs.

Looking Backward

This study represents work done over the period 1970 to
1976. It was virtually complete when I was asked to become
a member of President Carter's Council of Economic Advisers.
As a result, a few loose ends will appear, for chapters 3, 6,
and 7 were written on the fly. Basically, however, the work
represents a view of the world shortly after the oil crisis.

Looking backward, it seems useful to ask what seems right
and what seems wrong. I view the results described here as
primarily *methodological*, as indicating a technique by which
economic and engineering tools can be used to analyze trends
in markets and the contributions of new technologies. Since
the original version was written in 1973, this class of technique
has gained wide currency among energy modelers.

On the basic *empirical* results of this study, concerning the
efficiency price of energy resources, nothing has arisen to change
my view substantially. Undiscovered oil is being discovered
and is turning into proven reserves at a good clip; demand for
oil is growing at about what the energy demand functions
would predict; no major new findings or analytical problems
have been uncovered.

There are, however, three features of energy markets
which pose very great uncertainties and may modify the
empirical results, especially in the short run. These features are
the cost of new technologies, the determinants of OPEC

pricing, and the regulation of the economy. When recent evidence of these three features is combined, it appears likely that, especially in the short run, the cost and price trajectories outlined below are understated.

The *costs of new or unproven technologies* must perforce be laden with uncertainty. Until considerable experience is developed with technology, there is no way of predicting the cost of future technologies except with a large margin of error. Studies of cost overruns—essentially errors of cost estimates—in large construction projects and in military aircraft development indicate that the ultimate costs routinely vary by a factor of two from first estimates. At the same time, it is just as routine to note that after development and when engineers begin to move down the "learning curve," very rapid improvement in costs generally follows introduction.

Given these inherent uncertainties, I have particular concern that the cost of the first generation of new technologies —particularly synthetic fuels such as shale oil and gasified and liquefied coal—may have been substantially underestimated. The most recent studies of synthetic fuels indicate that costs are from 25 to 60 percent higher than the figures used in the present study. Assuming the most recent estimates are accurate and that no learning occurs after introduction, this would lead to higher costs, higher prices, and lower energy demands.

A second area in which there continue to be considerable uncertainties concerns *OPEC pricing* of oil and gas. In chapter 6 it is shown that there is substantial uncertainty about the wealth-maximizing price in the short run—the main uncertainty deriving from the uncertainty about the short-run price-elasticity of the demand for oil. In the long run, however, the different studies cluster fairly closely. There are no major new developments that would change these estimates; the most recent studies show no change from those given in chapter 6.

Many observers question the validity of such studies, however, because the models generally ignore the role of noneconomic factors in decisions about production and pricing. One problem is that the concept of maximizing net wealth is clearly too narrow an objective for many governments. Other important objectives are to preserve the "black gold" for future generations (to diversify the portfolios away from paper wealth) and to use the oil weapon for political ends. Some countries (Saudi Arabia, Kuwait, United Arab Emirates) are probably simply glutted with cash. There appear to be irrational elements in decision-making, in that discount rates placed on oil are different from rates placed on other goods.

Another set of concerns is the dynamics of cartel behavior. Countries whose reserves have but a short lifetime may put strong pressure, including military threats, to restrain production on countries with large reserves. There are, of course, pressures in the opposite direction. Many oil-producing countries are poor, and they will have a strong inclination to expand production as rapidly as possible as long as they are but a small fraction of the world market. These countries also generally have very high discount rates, which will tend to incline them to bring reserves to market quickly. All these and other factors make the use of the monopoly models suspect, indicating that the quantitative results are of limited validity.

A final area where the fundamental structure of the model can be questioned concerns the effect of *regulatory and environmental* policy. The fundamental assumption that runs through the analysis is that regulatory and environmental policies do not significantly impede market forces. Thus it is assumed that there is no further tightening of the environmental screws beyond levels of 1975. Further, it is assumed that (outside of OPEC) prices reflect costs of production and that there are no non-market impediments to developments or introduction of new technologies.

Each of these assumptions is probably too optimistic. The regulatory problems are perhaps best illustrated for environmental policy. Environmental policy in the energy sector in the United States presents flaws which lead to serious economic inefficiencies. The flaws stem from two basic areas. First, the technique of regulation for environmental spillovers relies on a series of detailed, specific technological requirements for individual firms and utilities as to how they should produce energy. As a result, we spread around our environmental dollars extremely inefficiently. A well-documented example is the $40-billion regulation proposed by the U.S. Environmental Protection Agency in 1978 to reduce sulfur emissions from coal-fired plants. This regulation mandates techniques be used to remove sulfur after mining in preference to using low sulfur coal. It does so on a nationwide basis. As a result, sulfur removal is inefficient—the nation pays more than needed for sulfur removal. More troubling is the possibility that overall health effects will be *worse* under the original, full-control proposal than under a well-designed standard. These and other inefficiencies are troubling because they may lead to costs that are two or three times what is efficient and to the stifling of new technologies.

The assumption of insignificant nonmarket impediments to energy production is also unrealistically optimistic. In the United States government regulation determines a significant fraction of price, allocation, and technical decisions. New technologies are largely funded by government. More troubling, however, is the increasing tendency to eliminate or slow development of energy technologies—like nuclear or coal—due to a "zero-risk" philosophy.

To a considerable extent, such inefficient and irrational policies make technological change and substitution uncertain or extremely expensive. If they become prevalent, there is little hope that we can make a gradual and inexpensive transition from oil and gas to alternative fuels.

I MARKET ALLOCATION
OF EXHAUSTIBLE RESOURCES OVER TIME

A universal problem faced by any economy is, how should its finite stock of scarce exhaustible resources be allocated over time? One answer to this question is to leave the allocation to the competitive market forces of the "invisible hand," where the interaction of profit-oriented firms and utility-maximizing consumers determines a time path. At the other extreme is the approach of the "visible hand," where the planning of government agencies or dominant firms influences or dictates the outcome. Whichever route is chosen, economists generally keep in mind the questions: How efficient is such an allocation? and, What can be done to improve the efficiency of the social choice mechanism that is used?

The present chapter focuses on the market allocation of exhaustible resources. It begins with an analysis of competitive markets—those in which both producers and consumers are "atomistic" and therefore do not influence the price through their own decisions. In addition, the chapter considers equilibria where there is a full set of futures markets, failing which the propositions about competition will not hold.[1] The final section briefly analyzes the effect of monopoly on resource allocation.

---

1. On the subject of resources, the literature is almost inexhaustible. The classic piece is Hotelling (1931). Of more recent statements, Solow (1974) is probably the best single source.

Before analyzing in detail the way competitive markets allocate resources, it is useful to highlight some of the efficiency properties of such markets. The major message that economic theory can bring to the discussion of the allocation of exhaustible resources is that *an internalized competitive equilibrium is efficient but not necessarily just*. This short sentence contains technical terms that need clarification before we proceed. First, what is meant by competition? As noted above, a competitive equilibrium is one where there are large numbers of individuals on both sides of transactions, with the result that no one individual can influence the price. In addition, there must be a full set of markets for different goods, regions, and time periods. In the United States the market forces have ranged from the "invisible hand" of competition in oil exploration and coal mining, through the oligopolistic structure of oil refining and distribution, to the heavily regulated pipelines and locally monopolistic electric utilities.

Second, what is meant by an internalized equilibrium? It is an equilibrium in which all the social costs and benefits are "internalized" to the relevant decision-maker. Allocational inefficiencies arise for goods with "externalities." When there are processes that generate significant volumes of pollution (such as coal burning) or services that yield unappropriated benefits to the future (such as learning by doing or invention), the presumption is that, unless internalized, the competitive market will produce too much pollution and too little learning.

Unfortunately, the income distribution generated by a competitive outcome may be unjust or unlovely from a moral or ethical point of view, but such an allocation cannot be economically wasteful in the sense that more useful goods could be produced in any period without either more inputs or fewer outputs in another period. In some ways, then, the outcome of the competitive markets is a standard against which performance can be measured; and in market economies, the inefficiencies that are associated with a particular allocation will be associated with imperfections or deviations of the market outcome from what a competitive market would provide.

Given this view of the way competitive markets function,

the analysis of the allocation of exhaustible resources is divided into a brief analysis of the allocation under competition in the first part of this chapter and a brief excursion into the analysis of imperfections in resource markets in the balance of this chapter.

A. *Competitive Equilibrium for a Single Grade of Resource*

The competitive allocation of resources will be examined in a dynamic partial equilibrium. Economy-wide variables such as wage rates, interest rates, and price of capital goods are taken as given. These, together with the technological know-how and resource availability, determine the cost functions for the resource. On the demand side, the demand functions show how quantity demanded varies as a function of the resource's price relative to other prices, of income, of population, and of other variables. Combining the technology and demand will yield a dynamic equilibrium path of prices and quantities.

More specifically, the analysis of exhaustible resources refers to the consumption of a given *stock* of resources. On the supply side, in general, the flow of the resource into the market is taken to be a function of the available stock, the market prices of the flow, and the extraction costs. Similarly, the demand for the resource is determined primarily by the price of the flow resource and income.

For a more formal analysis, assume that there are $\bar{R}$ recoverable units of a natural resource and that these are owned by competitive firms that attempt to maximize their net worth. At the start, assume that there is only one "grade" of the resource, that is, each unit of the resource costs the same amount to extract; and extraction cost as a function of time, $z(t)$, equals a constant z.

First consider the equilibrium condition for competitive firms. In analyzing the allocation it is crucial to distinguish between the gross price of the resource (e.g., the price of the barrel of oil) and the net price, or royalty (i.e., the price per barrel less the costs of production). In what follows, call the gross price "the price, $p(t)$," and the net price "the royalty, $y(t)$." The competitive firm must decide in what time period it should sell its resource. If $y(t)$ is the royalty in year t, then the present value of a sales plan $\{s^k(0), s^k(1), \ldots, s^k(T)\}$, where $s^k(t)$ is the total quantity sold in year t by the kth firm, is

$$V^k = y(0)s^k(0) + y(1)s^k(1)(1 + r)^{-1} + \ldots + y(T)s^k(T)(1 + r)^{-T}, \tag{1.1}$$

where r is the appropriate discount rate (here the interest rate) and V^k is the present value of the plan. The constraint for the producer is simply that

$$\sum_{j=1}^{T} s^k(j) \leqq \bar{R}^k, \tag{1.2}$$

where $\bar{R}^k$ is the quantity of the resource owned by the producer.[2] The present value of a unit sales in year t, $\rho(t)$, is the derviative of V^k with respect to $s^k(t)$, which in turn equals the *present value royalty*, $y(t)(1 + r)^{-t}$:

$$\rho(t) = \frac{dV^k}{ds^k(t)} = y(t)(1 + r)^{-t}.$$

Since V^k is linear in $s^k(t)$, the producer will produce in those period(s) when $\rho(t)$ is the highest, ρ^*. Put differently, if

$$\rho^* = \max[\rho(1), \ldots, \rho(T)], \tag{1.3}$$

then

$$\begin{aligned} s^k(t) &\geqq 0 \text{ if } \rho(t) = \rho^* \\ s^k(t) &= 0 \text{ if } \rho(t) < \rho^* \end{aligned} \tag{1.4}$$

If ρ^* is positive, present value maximization implies that the equality in equation (1.2) is binding and that the firm sells all its resource.

In a competitive market (ignoring transportation costs), the price of the resource is the same to all firms facing the same present value constraint. Thus industry sales in year t, $s(t)$, are given by

2. In this example, there are no "flow constraints" on production. In reality, most recovery decisions involve producing a stream of output over time rather than a quantity at a given point of time. We can, however, think of the sales as sales of the whole stream, as in a long-term contract for the output of a mine or well, and the royalty as pertaining to the stream.

$$s(t) = \sum_k s^k(t) \geqq 0 \text{ if } \rho(t) = \rho^*$$

$$s(t) = \sum_k s^k(t) = 0 \text{ if } \rho(t) < \rho^*. \tag{1.5}$$

A resource will be "scarce" if it commands a nonzero royalty. As noted above, as long as the royalty is positive, we know that

$$\sum_{k,j} s^k(j) = \bar{R}. \tag{1.6}$$

The equilibrium condition is shown graphically in figure 1.1, where continuous time (or short intervals) is assumed for the illustration. The present value of the royalty is at its maxi-

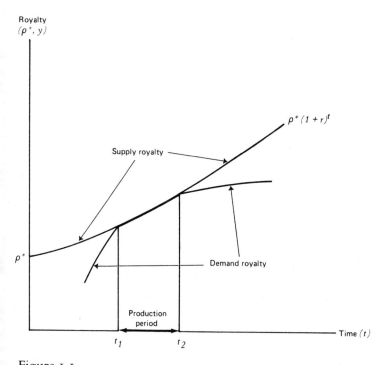

Figure 1.1.
Resource royalty in a competitive market equilibrium. Present value of royalty is maximized over period $[t_1, t_2]$. Production is zero over period before t_1 and after t_2.

mum, ρ^*, over the range $[t_1, t_2]$; only during this period does
production occur. The "supply royalty" shows the royalty
on the asset that must hold over the period until exhaustion to
clear the asset market—that is, the market for the resource-in-
the-ground. The demand royalty represents the royalty implicit in the market price (the demand royalty is market price
less extraction cost). The demand price always equals supply
price when production occurs. Before or after production
occurs, it lies below supply price.

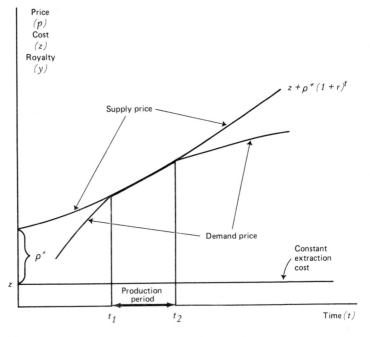

Figure 1.2.
Resource price in a competitive equilibrium. Supply price
equals extraction cost (z) plus compounded value of maximum
royalty $[\rho^*(1 + r)^t]$. Demand price is given by market equilibrium.

The price of the good, $p(t)$, will be given by the sum of royalty plus extraction cost during production periods. Figure 1.2 illustrates the competitive determination for the resource price, derived from figure 1.1 and the assumption of constant extraction cost.

To summarize, the basic results so far are as follows: First, the competitive equilibrium for exhaustible resources determines a royalty path for resources-in-the-ground that rises at the interest rate until the resource is exhausted. Second, as long as extraction costs are rising less rapidly than the interest rate, so will the product price of the resource.

A slightly different way of viewing the problem will be particularly helpful later. This is to look at the path in terms of the initial royalty and the exhaustion date. In figure 1.1 note that the use of the resource begins in period t_1 and ends in period t_2. By the reasoning given above, the resource grade examined is therefore exhausted exactly at point t_2. Shifting our vantage point to point t_1, we can ask two questions: What is the initial royalty $y(t_1)$? and, What is the date of exhaustion t_2? Obviously, the conditions for clearing the market are that the initial royalty be such that the resource be exhausted by t_2, or

$$y(t) = y(t-1)(1+r) \qquad (t \leq t_2) \qquad (1.7a)$$

$$\sum_{j=0}^{t_2} s(j) = \bar{R} \qquad \text{(if } y_0 > 0\text{).} \qquad (1.7b)$$

Equation (1.7a) is the *myopic condition* that, until exhaustion, the royalty on the resource must rise at the interest rate. Equation (1.7b) is the *global condition* that the consumption path induced by the price path be such that the resource is exhausted at date t_2.

As formal analysis is somewhat opaque, a simple numerical example is presented that shows how price behaves in a competitive resource market. Assume the resource is crude oil, with initial resources of $\bar{R}$ taken to be 1,000 billion barrels (1,000 BB or 140 billion metric tonnes). The time horizon is taken to be 40 years, after which a substitute in the form of a cheap synthetic liquid hydrocarbon, syncrude, will be found. The demand function takes the form

$$C = Ae^{gt}p^{-\eta},$$

where this can be met equally well either by crude oil or syncrude. C is consumption, $-\eta$ is the price elasticity of demand, t is time, and g is the exogenous growth rate in demand. Take the initial year to be 1970, a year in which actual consumption of oil was about 20 BB/year at a price of $4/barrel. Assume a growth rate $g = 0.05$, a price elasticity $-\eta = -1.0$, an interest rate $r = 0.10$. Therefore, using the 1970 price and quantity to determine the demand function, we find that it is $C(1970 + t) = 20\ exp[0.05t](p_t/4)^{-1}$. Given the rate of exhaustion, we want to determine the initial royalty that will satisfy the conditions in equations (1.7a) and (1.7b).

First examine the case where extraction costs are zero. In this case, price in a competitive allocation is rising at 10 percent per year, and equation (1.7b) becomes

$$1,000 = \int_{0}^{40} C(1970 + t)dt$$

$$= (20)\left(\frac{4}{p(0)}\right)\int_{0}^{40} e^{(0.05-0.10)t}dt.$$

The solution is that the price starts out at $1.38 per barrel and, compounding continuously, rises to $75 per barrel by 2010. Quantity is rising at $\left(g - \eta\frac{\dot{p}}{p}\right)$, or minus 5 percent per year.

B. *Equilibrium with Several Grades*

When there is more than one grade of resource, the analysis is somewhat more complex. Consider adding a second grade of resource, the extractions costs of the two grades being z_1 and z_2. For simplicity, assume that $z_1 < z_2$ but that both are constant over time.

The analysis about the price-output relationship analyzed in equations (1.1) through (1.7) and in figures 1.1 and 1.2 holds for each resource separately. This has an important implication in the simple model used here: the cheaper resource will always be used and exhausted before the second resource is touched.[3] The reason for this conclusion is shown in figure 1.3, where the production and price patterns for two kinds of resources are graphed. Consider competitive paths for the resources; each must satisfy the condition of equation (1.7a). The cheaper resource, with extraction costs z_1, has a supply price equal to the extraction cost plus the royalty, for example $p_1 = z_1 + y_1(0)(1 + r)^t$; the more expensive resource has price equal to $p_2 = z_2 + y_2(0)(1 + r)^t$. Each path must satisfy the condition that royalty is rising at the interest rate, and the path must meet the global condition of equation (1.7b) that supply is exhausted just when it no longer becomes economical. Thus the curve ABC, representing the time profile of the supply price of the first resource, leads to consumption over the period from 0 to t_1.

At point t_1 the exponential term on resource 1 means that the supply price on resource 1 is rising faster than the demand price ABFG, so at this point resource 1 becomes uneconomical and must be exhausted. Similarly, at time t_1 the supply price of resource 2 becomes equal to the demand price, and production of resource 2 begins. The history for resource 2 duplicates that for resource 1 in that its royalty must be such that it takes over just when resource 1 is exhausted and that its supply lasts just until the end of the period, that is, until exhaustion point t_2, at which point it becomes uneconomical. This remarkable teamwork between prices and quantities is a feature of the efficient allocation of resources. It will also appear in the optimizing programs investigated in chapters 5 through 8.[4]

It is easily seen from figure 1.3 that the cheaper resource must always be used before the second resource. This follows

3. For the more realistic model in which the analysis refers to drilling rather than to production (see footnote 2 of this chapter), the correct statement would be that low-cost resources are *drilled* before high-cost resources. Because it takes time to actually extract the resource, at any point of time extraction of different grades of resources may be occurring.

4. Aficionados of mathematical programming will recognize the property of "dual complementarity" in the teamwork of efficient prices and quantities.

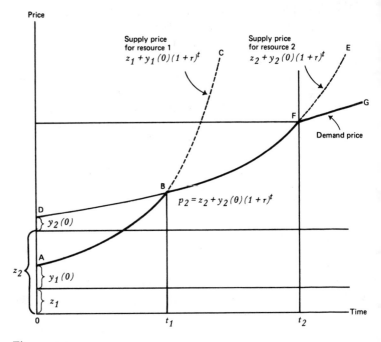

Figure 1.3.
Pricing pattern with two resources. Heavy line ABFG is price path of output. ABC is supply price of resource 1 with extraction cost z_1, while DBFE is supply price of resource 2 with extraction cost z_2. Dashed extensions of supply price lines indicate that these are purely accounting prices since resource is exhausted.

---

from the fact that in the efficient path the price of the first resource is always rising faster than that of the second resource because the royalty is larger. Thus if the second resource were used first—because its supply price were lower—the first resource would never be used.

C. *The Backstop Technology*

Up to now, I have paid scant attention to the transition problem: What happens when the economy "runs out" of its exhaustible resources? At one extreme of the possible treatments are the "doomsday" models that depict economic collapse. At the other extreme are optimistic models that assume a world of superabundant, low-cost synthetic or substitute resources. How can resource models treat the transition problem?

In the following analysis, a simplification is made that avoids complications associated with time horizons or global exhaustion. The simplification is the concept of a *backstop technology*. Today's energy technology is highly dependent on resources that are very cheap to extract but relatively scarce when viewed over a very long time horizon (drilled natural gas, Mideast oil). In this technology, royalties on scarce, low-cost resources may be relatively important in today's price. Over the next century or so, many low-cost energy resources will be largely exhausted, leaving more abundant but also more expensive resources (shale oil, coal). Ultimately, if and when the transition is completed to an economy based on resources present in superabundant[5] quantities—whether this be nuclear fission or fusion, solar, geothermal, or some as yet undiscovered technology—the economic importance of the scarcity of exhaustible resources will disappear, and capital and labor costs alone will determine prices. This ultimate technology—resting on a superabundant resource base—is the *backstop technology* and is central to this study of the allocation of scarce energy resources.

It is useful to illustrate the backstop technology by a simple example: Consider two processes for generating electricity. One process uses one unit of petroleum per unit output; petroleum resources are finite in supply ($\bar{R}$ recoverable units), free to extract, and can be converted to electricity at no cost. The second process uses sunlight, which is superabundant and free, and K dollars worth of capital per unit output. Assume the rate of interest is r, and further that demand is inelastic with D units of electricity demanded per year. It is clear from the earlier discussion that with a positive interest rate the petroleum

---

5. "Superabundance" in a technical sense means that a resource constraint, such as in equation (1.2), will never be binding, so that the shadow price is always zero on the "superabundant" resource.

resources will be used first, and that the switch will be made to the solar process (the backstop technology) at $\hat{T} = \bar{R}/D$ years in the future.

Prices along an efficient path are easy to calculate. At the switch point $\hat{T}$, and forever thereafter, the price per unit electricity, p, is given by the backstop cost:

$$p(\hat{T}) = (r + \delta)K,$$

where δ is the depreciation rate on capital. From our earlier discussion, this implies that the price, and therefore the royalty on petroleum at the switch point, is also $p(\hat{T})$. But then price and royalty along the efficient path until $\hat{T}$ are simply

$$y(t) = p(t) = p(\hat{T})e^{-r(\hat{T}-t)} = (r + \delta)Ke^{-r(\hat{T}-t)} \qquad (1.8)$$

The royalty on the scarce resource is simply the switch price, $p(\hat{T})$, discounted back to the present.

The price path is illustrated in figure 1.4. The three important elements in determining current royalty are: the capital cost of the backstop technology, the interest rate, and the switch

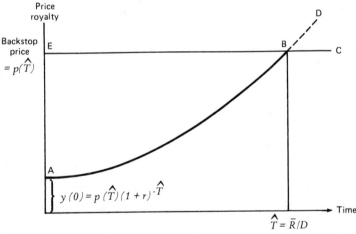

Figure 1.4.
Pricing of an exhaustible resource with a backstop technology. ABD is the supply price of the scarce resource, while EBC is the supply price for the backstop technology. ABC is the market price of the energy good.

date. The capital cost of the backstop technology (K) enters linearly. The interest rate enters positively as a linear function of the cost of the backstop technology and negatively as a discount factor applying to the switch date. For fixed $\hat{T}$, a higher interest rate raises $p(0)$ if $\hat{T}(r + \delta) < 1$ and lowers $p(0)$ if $\hat{T}(r + \delta) > 1$. The switch date $\hat{T}$ enters in an exponential way in much the same way as the interest rate. In our simple example without growth, recall that $\hat{T} = \bar{R}/D$. If the amount of resources doubles, or demand halves, the switch date is doubled. This lowers the royalty by the factor $exp[-r\bar{R}/D]$.

An important extension is the introduction of techno-logical change. Assume for simplicity that technological change is proceeding at rate λ throughout the energy sector relative to the rest of the economy. This affects the *costs* of extraction and of the backstop technology, but *not the growth rate of the royalty*. Thus $z(t) = z_0 exp(-\lambda t)$ and $K(\hat{T}) = K_0 exp(-\lambda \hat{T})$. Returning to the simple example in equation (1.8), the equation is modified by technological change to yield the following:

$$p(t) = z_0 e^{-\lambda t} + [(r + \delta)K_0 e^{-\lambda \hat{T}} - z_0 e^{-\lambda \hat{T}}]e^{-r(\hat{T} - t)}$$
$$= z_0 e^{-\lambda t} + \gamma(\hat{T})e^{-r(\hat{T} - t)} \qquad (1.9)$$

and royalty is

$$\gamma(t) = [(r + \delta)K_0 - z_0]e^{-(r + \lambda)\hat{T} + rt}. \qquad (1.10)$$

Thus technological change has the same effect on the royalty as increasing the discount rate from the switch date (although royalties still rise at the interest rate). In addition, technological change gives a downward tilt to the price path, relative to no technological change, because the extraction costs are declining over time.

This oversimplified example of the allocation with a back-stop technology will form the basis for the empirical estimates of efficient energy prices in chapters 5 through 8 of this study. The question I will attempt to answer is whether the current market-determined royalty on energy resources appears to be close to that which would come out of an efficient path for the allocation of energy resources. As is indicated by the equation for price, if the price of the backstop technology is low—or if the switch date is far off or if the interest rate is high—then the royalty on energy resources is relatively low. Conversely, if these are reversed, the royalty on energy resources is high. Unfortunately, the calculation required to obtain the answer is extremely complex. There are many sources and grades of energy resources, many uses, and many demand categories, each with peculiar specifications. Switch points will differ, so calculation of the optimal path and the switch points for different resources is cumbersome. Nevertheless, the basic picture depicted in simple models remains.

D. *Market Power in Resource Markets*

Historically, the energy market has been the centerpiece in the genesis and controversy in imperfect competition. From the time when John D. Rockefeller and the Standard Oil Company discovered the profitability of concentrating ownership, there has been a pitched battle between the companies and the public over the extent of monopoly power and the abuses of oil companies. Until about 1970, the Western oil companies were predominant in the production and distribution of petroleum products. With the dismantling of colonial empires and the rise of nationalism among the producing countries outside of the developed West, the national governments themselves replaced the oil companies and largely perfected the techniques of monopolization by forming an international cartel (the Organization of Petroleum Exporting Countries— OPEC). What was once a minor conflict between producers and consumers on a national level is today projected on the international screen as a major political and economic conflict between producing and consuming countries.

It is commonplace to note that oil is but one of many possible commodities where producer cartels could exert significant market power. On the other hand, the conditions for the exercise of market power are better met in the petroleum sector than in other commodities: OPEC oil deposits are far cheaper to extract than oil deposits in other areas. These deposits were monopolized in earlier times by the oil companies; currently they are monopolized by the national governments. Strictly speaking, however, since there are many producers of these low cost deposits, OPEC production is oligopolistic.

The degree of concentration in the petroleum and the energy market is shown in table 1.1. This table shows four-unit and eight-unit concentration ratios for petroleum—by exports, production, and reserves—as well as the shares for fossil fuels as a whole. In this treatment, the national governments are considered as "units." In principle, it makes little difference whether the market is controlled by four big firms or by four big governments. Clearly, the market for petroleum, especially petroleum exports, is today highly concentrated. Reserves are marginally less concentrated, and the production in the U.S. energy market is moderately competitive by the usual standards. The figures suggest that the problem of market power in

Table 1.1.

Concentration in Energy Markets, Recent Periods (By percentage)

	Top four units		Top eight units	
	U.S. market*	World market	U.S. market*	World market
Petroleum				
Exports	n.a.[†]	54	n.a.[†]	78
Production	31	56	49	73
Proved reserves	33	53[‡]	55	73[‡]
All fossil fuels				
Exports	n.a.	46	n.a.	67
Production	19	53	32	67
Reserves	40	n.a.	60	n.a.

SOURCES: World production and export data from United Nations (1976), *Energy Facts* (1975), and *Oil and Gas Journal*, December 1976. U.S. data from FTC (1974).

NOTE: Figures are for 1973 unless otherwise noted. Outside the U.S. a unit is defined as a sovereign government; in the U.S. market a unit is a firm. All resources are converted to equivalent thermal inputs. Fossil fuels include natural gas, coal, and petroleum.

* 1970.

[†] Not available.

[‡] End 1976.

petroleum, especially exports, is presently quite significant. In the future, as the locus of the petroleum production broadens to other countries, and as high prices induce interfuel substitution (coal, shale, nuclear, etc.), the *potential* for exercise of market power will narrow.

In examining the exercise of market power, one serious analytical difficulty arises. As shown above, there are royalties, or shadow prices, for most energy resources that arise due to scarcity. Unfortunately, there is no way of determining a priori whether the difference between the market price and the marginal cost of production is the appropriate competitive royalty or monopoly profits. This difficulty is one of the most important questions addressed in the empirical part of this study. Thus from 1970 to 1974 the payments to producing countries in the Mideast went from $1 per barrel to $9 per barrel. It is not possible from an analytical point of view to say whether the move was due to exercise of monopoly power or to a recomputation of the competitive royalty in light of new information.

The problem of exercise of monopoly power in resource markets is a straightforward application of standard monopoly theory.[6] If there is no exhaustible resource, then the standard model applies exactly. When either the monopolist or his competition has only a finite resource base—as in the case of petroleum—then the intertemporal allocation problem enters through the royalty on the scarce resource. Although there are many possible configurations, the most plausible starting point is to assume that there is a competitive *backstop technology* (R_2) and an *exhaustible technology* that is monopolized (R_1). In this case, we assume that the supply of competitively produced R_2 is perfectly elastic at a price $p = p^*$; while the supply of R_1 is determined by a monopolist with marginal cost $c < p^*$. In addition, assume that producers set prices to maximize wealth (or the discounted value of net revenues). The final important assumption at this stage is that the consumers do

---

6. The basic elements of monopoly theory are outlined in Baumol (1971). The early analysis of monopoly in resource markets is found in Hotelling (1931).

not retaliate, and producers expect them not to retaliate, against the exercise of monopoly power.

More precisely, consider a market where the demand for output of the resource is given by an inverse demand function $p_t = f_t(Q_t)$, with revenue function $R_t = Q_t p_t = Q_t f_t(Q_t)$. Profits per period are then $\pi_t(Q_t) = Q_t(p_t - c) = Q_t[f_t(Q_t) - c]$. The discounted value of profits is then

$$W = \sum_{t=0}^{\infty} Q_t[f_t(Q_t) - c](1 + r)^{-t} = \Sigma \pi_t(Q_t)(1 + r)^{-t},$$
(1.11)

where r is the monopolist's discount rate on profits. This will be maximized subject to the constraint that total production is no greater than monopolist reserves (R) or $\Sigma Q_t \leqq R$. Putting this together results in the Lagrangean expression:

$$W = \Sigma \pi_t(Q_t)(1 + r)^{-t} - \lambda[Q_1 + \ldots + Q_T + \ldots - R].$$

First order Kuhn-Tucker conditions for a maximum are:

$$\pi_t'(Q_t) \begin{cases} = \lambda(1 + r)^t \text{ for } Q_t > 0 \\ \leqq \lambda(1 + r)^t \text{ for } Q_t = 0. \end{cases}$$
(1.12)

Equation (1.12) states that through the period of monopoly production the *marginal royalty* $[\pi'(Q)]$ *must rise at the interest rate*. This differs from the competitive condition, which states that the *average* royalty $[\pi(Q)/Q]$ rises at the interest rate.

This condition is most easily interpreted geometrically. In what follows it is assumed for simplicity of exposition that the demand curve and cost functions do not change over time. Figure 1.5 shows the demand curve for the product and the marginal revenue curve for the product (both excluding the backstop technology) as HCE and JFG, respectively. However, since the monopolist must reckon on the substitution of the backstop technology when the price rises to p^*, the monopolist demand and marginal revenue curves are ABCE and ABCFG, respectively. The monopolist curves, labeled DM and MRM, are the product demand and marginal revenue curves cut off at the backstop price.

The condition for the price path to maximize wealth given in equation (1.12) is that the marginal royalty, $\pi' = \text{MRM} - c$,

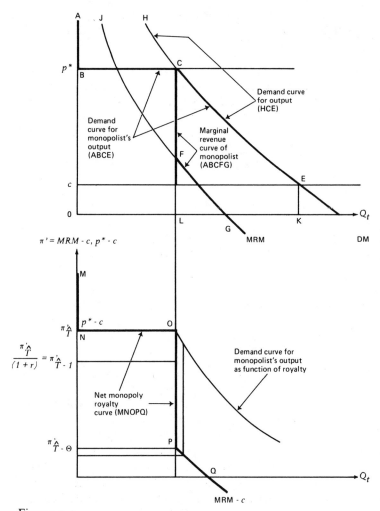

Figure 1.5.
Demand curves and royalty curves shown as function of monopoly production (Q_t). Top diagram shows demand curve for output (HCE); demand curve for monopoly output (ABCE); marginal revenue curve for monopolist (ABCFG); and extraction cost (c). Bottom diagram shows determination of monopolist's optimal royalty. Net monopoly royalty curve (MNOPQ) and discounted net royalty determine optimal pattern of production.

will rise at the interest rate. In addition, the monopolist must exhaust the resource, since if he did not he could make additional profits and the path would not have been optimal. Thus the price path is such that the marginal royalty rises at the interest rate and is greater than $p^* - c$ after the period in which the resource is exhausted (period $\hat{T}$).

The pattern of marginal royalties is shown in the bottom half of figure 1.5. We know that the marginal royalty in the last period is (just under) $p^* - c$, shown as $\pi_{\hat{T}}'$ in the figure. By discounting this back, we obtain the marginal royalty in each earlier period; and by seeing where the π_t' line cuts the $(MRM - c)$ curve, we can determine the optimal output for the monopolist. The important point to note is that, for a considerable range in the example shown in figure 1.5, the marginal royalty curve cuts the $(MRM - c)$ curve on the vertical section, implying that the price is at the backstop price. This outcome reflects the fact that at the backstop price the marginal revenue is sharply lower because of the "kink" in the demand curve for the monopolist at point C in the top part of figure 1.5.

Thus for θ periods before the exhaustion of monopoly resources, the marginal royalty is on the vertical portion of the net monopoly royalty curve. Going back $(\theta + 1)$ periods, however, takes the marginal royalty below point P in figure 1.5, and monopolist's output will be higher. Only in period $T - \theta - 1$ or before, then, is the monopolist's price lower than the backstop price.

The following concrete example illustrates the point. Assume that there are no extraction costs, that the backstop price is \$10, that the discontinuity in the demand curve extends to \$1, and that the interest rate is 10 percent. In this case, the monopoly price will be at the backstop price for a period of 24 years before monopoly resource exhaustion.[7] Only for the period more than 24 years before the exhaustion date will the

---

7. The formula for the point of discontinuity is easily determined. If the demand curve has elasticity $-\eta$, the marginal royalty is $\pi'(Q) = (1 - 1/\eta)p(Q) - c$. Thus if the backstop price is $p^* = 10$, $c = 1$, and $r = 0.10$, then for $\eta = 2$ the discontinuity runs from $p^* - c = 9$ to 4 and $\theta = 8$; for $\eta = 4$ the discontinuity runs from 9 to 6.5 and $\theta = 3$.

price be on the sloping part of the monopolist's demand curve. Further, if the cost, c, is close to the backstop price, or if demand has a price elasticity less than one, then the marginal royalty will be such that the backstop price is always charged. The solution where the backstop price is charged is called the *limit price solution*, indicating that the price is limited by the competitive backstop price.

One additional result about these paths is of interest. It is easily seen that the price path for the monopolist lies above the price path for a competitive industry until late in the price path. The comparative price paths are shown graphically in figure 1.6, where an elastic case is displayed. Only when the competitive path gets quite close to exhaustion do the two paths cross. Given this shape of the price paths, it is clear that the path for the monopolist has a later exhaustion date than that for the exhaustible resource, since the price path is above the path for the competitive path and the quantities consumed in each period are therefore smaller for the monopolist up to the date of exhaustion of the competitive path. This result is the basis of Robert Solow's paradox that the monopolist is the conservationist's best friend.

It is useful to get a rough indication of the extent to which the monopolist delays the date of exhaustion by his restrictive policy. For the case where demand is unit elastic (price elasticity equal to minus one), with a discount rate of 10 percent, a growth rate of 2 percent annually in the demand curve, and the reserves equal to 50 years of demand at the backstop price, the reserves would last about 17 years under competitive management and twice that, 35 years, under monopolistic management. The details of the solution are quite interesting because *in the inelastic range of the solution the monopolist behaves as if he were a competitor with a zero interest rate.*[8]

8. This conclusion comes from solving the competitive and the monopolist paths where demand is given by $Q = p^{-\eta}e^{gt}$, $\eta = 1$, $g = 0.02$; further $p^* = 10$, $R = \int_0^{\hat{T}} Q = 5$, and $r = 0.10$. The relevant equations for the exhaustion date, are $Rp^*(g - r) = exp(g\hat{T}) - exp(r\hat{T})$ for competition and $Rp^*g = exp(g\hat{T}) - 1$ for the monopolistic case.

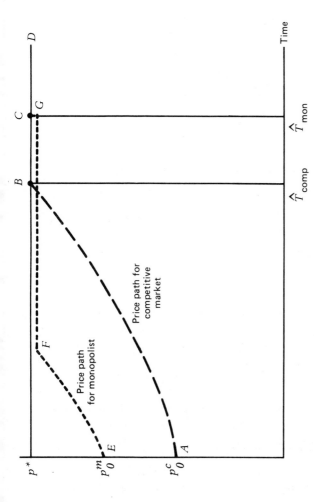

Figure 1.6.
Price paths for monopoly and competition. The path ABCD shows the price path for competition, with exhaustion at $\hat{T}_{comp}$, while the price path EFGCD shows the price path for a monopolist, with exhaustion at $\hat{T}_{mon}$. The path for monopolist is with elastic demand curve, which implies that for some time the path lies below the price of the alternative technology. Further note that the exhaustion date for the monopolist lies later than the exhaustion date for the competitive path.

2 THE DEMAND FOR ENERGY

A fundamental building block of the energy market, and of the energy model used in this book, is the demand for energy. This chapter outlines the relation between preference functions and demand functions and then presents the econometric estimates used in this study.[1]

Aside from being a building block in the model to follow, energy demand is an important issue in its own right. The central uncertainties in energy demand are four: First, as far as the long run, the central question concerns the income elasticity of the demand for energy, that is, for a given rate of growth in the aggregate output of an economy, what is the fractional increase of the demand for energy? Second, and relating mainly to the medium term, what is the long-run price elasticity of the demand for energy? Given the dramatic changes in the price of energy relative to other goods, it is of central importance to know what the eventual response of energy demand will be, especially the response of energy demand to further price-oriented policy measures. Third, for the short run the critical question is: What is the time distribution of the response to the recent price increases? Fourth, it is implicit in the questions usually raised that energy is a conventional economic good in the sense that it responds to relative prices, income, population, and other determinants in the same way that most other goods do. Many have argued

1. For a complete report on the demand study, see Nordhaus (1977).

22

that energy is unique—some that it is the ultimate determinant of value—and that the behavior of the demand for energy cannot be explained with conventional econometric models. Is this so?

A. *Theoretical Issues*

The goal of the estimation of demand relations is to determine the social valuation of energy products revealed in market activity. The fundamental relations underlying social valuation are individual *preference functions* and their aggregation into (possibly badly behaved) aggregates. It is assumed that society's preferences can be represented as a well-behaved preference function of the final goods and services consumed. This can be derived either from market demand functions for decentralized economies or sectors or from the preferences of the planners or representatives in a centralized economy or sector. Finally, it is presumed that the agents of the economy act, at least in the long run, to attain the most preferred set of goods. It is central to the view presented here that consumers desire energy *services* (passenger miles, warm rooms, haircuts) while energy *products* (gasoline, oil in furnaces, electric clippers) are means of fulfilling these desires.

To illustrate the preference relation, figure 2.1 shows a set of indifference curves between energy services (E) and nonenergy goods (X). Each U_i curve represents those points along which consumers are indifferent between different bundles; higher numbered indifference curves are preferred to lower numbered curves.

Consumers are faced with budget constraint $Y = p_E E + X$, where Y = income in terms of nonenergy goods and p_E is the relative price of energy to nonenergy goods. Faced with a budget constraint ABY_1 in figure 2.1, with slope p_E and income Y_1, a utility-maximizing consumer will choose point B.

The preference function cannot be observed directly, but in principle it can be reconstructed from observed data. In the real world, we can observe how consumers respond to different prices and incomes. Thus every time a consumer is faced with a price-income pair (p_E, Y), he makes a choice of (E, X). Thus in figure 2.1, point c_1 represents the point given

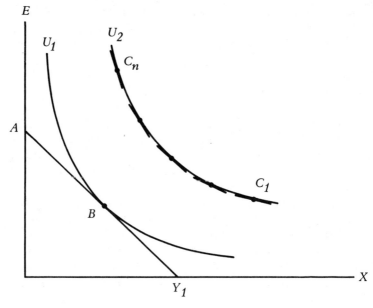

Figure 2.1.
Preference function over energy goods (E) and nonenergy
goods (X) shown by U_1, U_2, ...; budget constraint with
income Y_1 shown by ABY_1; consumer chooses B. Graphical
reconstruction of indifference curve U_2 from observed price
and income data is illustrated.

———————

by the quantity choice (E, X), and the little line through c_1 is
given by the budget constraint. A large number of such
hypothetical segments representing different experiments are
drawn along indifference curve U_2. In principle, all the little
segments can be joined together or integrated to reconstruct
the preference functions completely.[2] Thus the segments
$[c_1, \ldots, c_n]$—each representing a chosen point when faced by a
particular budget constraint—can be integrated to determine
indifference curve U_2.

In practice, the complete integration of observed points

———————

2. This is subject to integrability conditions; see Samuelson (1948).

is impossible because of an insufficient number of observations. However, an approximation to the preference function can be obtained by: (1) specifying the functional form of the preference function in terms of a parametric function; and (2) estimating the parametric preference function with econometric techniques.

In what follows, I use a very special representation, one that assumes that the preference function is separable over time. The notation is that c_t is an index of consumption, taking nonenergy goods as numeraire, and is given by:

$$c_t = X_t + \sum_{i=1}^{n} A_i E_{it}^{-\alpha_i} (X_t + \sum_{i=1}^{n} w_i E_{it})^{\beta_i} \qquad (2.1)$$

where X_t is an index of nonenergy goods; E_i are different energy services; A_i, α_i, and β_i are parameters of the utility function. The w_i are taken to be the base period weights for energy in terms of other goods, so that the expression in parentheses in (2.1) is Y_t. It is recognized that equation (2.1) is a restrictive formulation, but it is flexible enough to determine the important properties of a general utility function.

Maximizing equation (2.1) subject to the budget constraint $\Sigma p_{it} E_{it} + X_t = Y_t$ (p_{it} being current year energy prices of energy services relative to nonenergy prices) yields the following energy demand functions:

$$E_{it} = k_i p_{it}^{-\gamma_i} Y_t^{\delta_i}, \, i = 1, \ldots, N \qquad (2.2)$$

where $k_i = (\alpha_i A_i)^{1/(1-\alpha_i)}$, $\gamma_i = (1 - \alpha_i)^{-1}$, $\delta_i = \beta_i (1 - \alpha_i)^{-1}$.

B. *The Data and Variables*

In what follows, energy services are defined as total net energy consumed in a sector; the composition of total consumption between fuels is ignored. The important difference between this and earlier studies is that this study considers the demand for delivered (or net) energy whereas most other studies focus on gross energy. The concept of delivered energy was introduced by Hoffman (1972) and is central to the programming models of the energy sector [see Nordhaus (1977) for a fuller discussion].

Gross Domestic (or National) Product is taken to be the

aggregate income measure, and the aggregate price index is the GNP or GDP deflator. Per capita variables refer to total population. The time period for the study was generally from 1955 through 1972.

Electricity was aggregated with other fuels in each sector's consumption. In some cases, this aggregation satisfies the aggregation condition discussed in Nordhaus (1977). Thus electricity may be used for residential heating or cooking because it is the least costly in a given area. In other circumstances, electricity is used because the demand is for *specific electric* use; lighting and electric drive motors are areas where use of fuels other than electricity would generally be out of the question.

An ideal solution would be to divide electricity into general and specific electrical uses and to estimate these separately. In practice, such a separation is empirically difficult. There are estimates that the share of electricity for specific electricity in the United States for recent years is 50 percent in residential and commercial, about 100 percent in industry, and zero for transportation.[3] Unfortunately, there are no time series for all countries to make the separation. On the other hand, the specific electrical category cannot be simply ignored, for this would rule out use of electricity in early years of an efficient program.

In order to obviate this difficulty, I have assumed that a certain fraction of the demand in each category is specific electric. Following estimates from Beller (1975), the figure is estimated to be 20 percent of net residential demand, 10 percent of net industrial demand, and zero percent of net transport demand. These constraints are imposed upon the demand pattern in the linear programming solution. For the exact elasticities used in the program, as well as the methodology for deriving them, see section D and table 2.3 of this chapter.

The sample and methods are more completely described in Nordhaus (1977). The sample consists of observations on energy consumption, price, and output in seven industrialized countries: Belgium, Federal Republic of Germany, France,

3. See Beller (1975).

Italy, Netherlands, United Kingdom, and United States. The
time period of the study covers, for most countries, 1955 to
1972, for a total of 82 observations for each sector. There are
four sectors in the study—industrial, transportation, residential,
and energy.

C. *Econometric Results*

In the results that follow, I have presented estimates only
for the pooled sample. In the pooling of individual countries,
it is assumed that countries have the same preference functions
and production functions, except for scale factors. Since the
rate and level of technological change drop out of the equation,
there is no need to assume these to be the same across countries.
The major difficulty in pooling countries revolves around the
question of the appropriate conversions between different
currencies.

The usual procedure is to use market exchange rates, but
these are seriously deficient. First, it is clear that for market
economies, market exchange rates reflect, in part, volatile
temporary factors, and that temporary movements do not
reflect genuine changes in the relative real incomes of different
countries. A superior way of measuring real incomes is to use
"purchasing power parity" exchange rates, ones that compare
the purchasing power of incomes in different countries. These
are used to translate each currency into a "universal" standard
of value for a given year, while domestic output deflators are
then used to indicate changes over time.[4]

C.1 *Results for the Pooled Sample with Country Effects*

Only results for the pooled sample are presented as these
are what are used in the optimization model.[5] The theoretical
basis for pooling countries is to assume that all countries have
similar preference functions and production functions, but
that the differences in incomes and relative prices lead to dif-

4. This procedure is discussed by Balassa (1964). The purchasing power
parities used in this study are drawn from Balassa.

5. The results for the unpooled estimates, presented in Nordhaus (1977)
but not given here, are summarized in table 2.2.

ferent energy-intensiveness in different sectors. Thus we would expect that with high gasoline prices and low incomes in Europe, the amount of gasoline consumed in Europe per person would be considerably below that in North America, which has low relative gasoline prices and high relative incomes. In addition to the systematic effects of prices and incomes, there may be other omitted variables that are crucial to the determination of energy demand. The weather is clearly important in determining domestic heating demands; the road network in determining automotive demand; the industrial structure in determining the industrial demand. I have assumed that these effects, which can be called *country effects*, are multiplicative and do not vary systematically *over time*. This implies that we can use country dummy variables in our logarithmic specification to represent the effects for individual countries. I would be surprised if these country effects were nil; on the other hand, I would be disappointed if they accounted for too much of the variation. Thus the specification for the pooled model is that countries have different *levels* of energy demand, but that the elasticities, or response to prices and incomes, are constrained to be the same.

To construct the equations, I made the following simplifications. First, the current and lagged income terms appeared to have the same sized coefficients, so I constrained them to be equal. Next, I assumed that the lag on prices was linear over a five-year period; this lag is undoubtedly too short, but the shape is probably roughly correct. With these assumptions I reduced the equations to the following:

$$q_{t,i,j} = \alpha_{i,j} + \beta_j \left[\sum_{\theta=0}^{4} 0.2 p_{t-\theta,i,j} \right] +$$
$$\gamma_j [0.5 y_{t,i,j} + 0.5 y_{t-1,i,j}] \tag{2.3}$$

where i = country,
j = sector,
q = log of per capita net energy consumption,
p = log of relative net price of energy,
y = log of per capita real GNP,
$\alpha_{i,j}$ = individual country effects,
β = the *common* long-run price elasticity,
γ = the *common* long-run income elasticity.

Table 2.1.
Results of Energy Demand Estimates, Pooled Equation

Sector	Price (β)	Income (γ)	$\bar{R}^2$	SEE	$\overline{DW}$
Aggregate	−0.85 (0.10)	0.79 (0.08)	0.988	0.049	0.74
Transportation	−0.36 (0.12)	1.34 (0.08)	0.994	0.047	0.55
Residential	−0.79 (0.08)	1.08 (0.12)	0.990	0.059	1.01
Industry other than energy	−0.52 (0.17)	0.76 (0.16)	0.947	0.091	0.56
Energy	−0.58 (0.11)	−0.05 (0.12)	0.979	0.079	0.66

NOTE:
Dependent variable = per capita net energy in sector.
β = price elasticity of demand
γ = income elasticity of demand
$\bar{R}^2$ = corrected R^2
SEE = standard error of estimate
$\overline{DW}$ = Durbur-Watson statistic corrected for gaps in data.

Figures in parenthesis are standard errors of coefficients.

The results for the four individual sectors and the aggregate are summarized in table 2.1. In what follows, I focus on the long-run price and income elasticities.

In all four demand sectors, the price elasticities have the theoretically predicted sign and are well-determined while all the income elasticities are very well-determined.[6] First, concentrating on the price elasticities, it is seen that these are −0.36 (±0.12) for the transportation sector, −0.79 (±0.08) for the residential sector, −0.52 (±0.17) for industry other than energy, and −0.58 (±0.11) for the energy sector. In the aggregate, the estimate is −0.85 (±0.10). These results are not out of line with results of other studies.[7] Elasticities of

6. The text gives the estimated coefficients plus or minus the estimated standard error of the coefficient.
7. See Taylor (1977).

this magnitude indicate that *the long-run response of energy consumption to price is very substantial.* To the extent that the differences between the coefficients are significant, they indicate that the demand for energy is most inelastic in the transportation sector, followed by intermediate values for industry other than energy and energy, and that the residential sector is most elastic. The relative inelasticity of the transportation sector is quite plausible, since there is probably the least possibility for technological substitution in this field. On the other hand, the relatively high elasticity of the residential sector is plausible because of the high degree of substitutability between fuels and capital in that sector.

The results for the income terms are quite striking. The income elasticities are 1.34 ($\pm$0.08) for the transportation sector, 1.08 ($\pm$0.12) for the residential sector, 0.76 ($\pm$0.16) for the industry other than energy sector, and $-$0.05 ($\pm$0.12) for the energy sector. For the aggregate, the elasticity is estimated to be 0.79 ($\pm$0.08). Again the income elasticities are plausible from an a priori point of view. It is well known that private automobiles are both highly income elastic and relatively energy-intensive, so that the high income elasticity of transportation is not surprising.

In considering these results, three important differences from other studies should be noted: First, the results are found by pooling seven countries. As can be seen by comparing with the results from individual countries in Nordhaus (1977), results with unpooled data show no resemblance to the pooled results. Second, the concept of energy consumption is *net* energy, whereas most other studies for sectors examine gross energy. Since the general trend has been toward more efficient resources (natural gas and electricity as compared to coal), this leads to a more rapid growth of net energy. Third, the demands are for the entire sector rather than a single resource (e.g., electricity or natural gas) in a sector or an economy.

C.2 *Results for the Pooled Sample without Country Effects*
There is a troubling lack of elegance about the use of dummy variables: they are admissions that the specifications are incomplete. In addition, they may throw away consider-

able information about the effects of *international* differences in prices and incomes on international differences in energy intensiveness. For this reason, it is useful to perform these calculations without dummy variables. This procedure then takes into account not only the effect of the histories of individual countries but also the differences of levels of income and price between countries on energy intensiveness. To obtain this different perspective, we must make the further heroic assumption that the intercepts in all countries are the same and that omitted variables are uncorrelated with energy prices and income. The quality of the fit will deteriorate if country dummies are significant, but the results may shed further light on the long-run elasticities.

Table 2.2 summarizes the results for (1) the estimates for individual countries aggregated in a composite statistic;[8] (2) the results of the pooled equations shown in table 2.1; and (3) the results of equations that are pooled time-series cross-section and in which there are *no* individual dummy variables for different countries. Two results are clear from this table: first, the results generally hold up without country variables. Second, the fits of the equation are much worse. In considering the two equations, there are good theoretical reasons to believe that the results without country variables should show larger price coefficients: in principle, the price differences are of longer duration, and the full response to these differences should have taken place. For the time-series analysis of individual countries, the length of response is only five years, which is clearly too short for the response function for energy.

It does not appear that the results are significantly different with two exceptions: (1) in the transportation sector the price elasticity is much higher whereas the income elasticity is lower; and (2) in the energy sector the income elasticity is dramatically changed. With these exceptions, these results confirm quite strongly the results with the dummy variables.

The question is how to interpret the two cases where the results are quite different. In general, I suspect that for trans-

8. These results are presented in Nordhaus (1977), but for brevity are only summarized here.

Table 2.2.
Comparison of Estimates for Price and Income Elasticities with and without Country Dummy Variables

Sector	Price elasticities			Income elasticities			Goodness of fit (R^2)	
	Composite statistics	With dummies	Without dummies	Composite statistics	With dummies	Without dummies	With dummies	Without dummies
Aggregate	-0.66 (0.26)	-0.85 (0.10)	-1.15 (0.10)	0.84 (0.11)	0.79 (0.08)	0.87 (0.09)	0.988	0.916
Transportation	-0.36 (0.22)	-0.36 (0.12)	-1.28 (0.06)	1.68 (0.10)	1.34 (0.08)	0.81 (0.08)	0.994	0.959
Residential	-1.14 (0.29)	-0.79 (0.08)	-0.71 (0.09)	0.44 (0.17)	1.08 (0.12)	1.39 (0.12)	0.990	0.857
Industry other than energy	-0.30 (0.23)	-0.52 (0.17)	-0.48 (0.14)	0.78 (0.17)	0.76 (0.16)	0.91 (0.14)	0.947	0.671
Energy	-0.33 (0.25)	-0.58 (0.11)	-0.62 (0.17)	0.18 (0.14)	-0.05 (0.12)	0.94 (0.23)	0.979	0.606

portation the pooled results without country dummies should be given considerable attention. In this sector the differences between countries are quite clearly due to the policy of taxing gasoline heavily in European countries, rather than supply side differences. For this reason, results without country dummy variables probably are measuring a longer run reaction than results with country dummies, and therefore these seem to be more adequate results from a theoretical point of view. For the energy sector, on the other hand, the differences are probably supply side differences, in particular differences in energy resource availability, rather than demand. Clearly, the reason the demand in the energy sector is high for the United States, the United Kingdom, and the Federal Republic of Germany and low for Italy and France is due to the resource endowments of the respective countries. In fact, the causality may be reversed: countries that have larger energy sectors may therefore have higher incomes. For this reason, in the energy sector the results with country dummies are preferable to the results without country dummies.

D. *Transformation of the Econometric Estimates for Use in Programming Model*

In the programming model, total energy demands are projected for each of four consuming sectors. The demand for energy in each sector is measured at the point of end use, taking into account end-use conversion losses (and hence is equal to the energy delivered to the sector times the relative efficiency of the given fuel in that sector). Three traditional consuming sectors are represented: residential/commercial, industrial, and transportation. Because many end-uses (e.g., air conditioning, electric appliances, and electric motors) can only be satisfied by electricity if one is to avoid very sizable losses in economic efficiency, a fourth demand category, termed "specific electric," is also included. A constraint of the programming model then prevents the specific-electric demand of any sector from being satisfied with nonelectric energy.

A problem arises because the data used for the econometric estimates presented in this chapter did not allow the separation of specific electric from general electric. It was therefore necessary to impose an arbitrary rule for estimating the elasticity of specific electric energy demand, while at the same time ensuring the consistency of that elasticity with the overall elasticities. For this purpose the "specific-electric" elasticities with respect to both income and price were calculated by assuming that the "specific-electric" portion of each sector's demand (which is by assumption independent of the rest of that sector's demand) has an elasticity identical to that for the rest of the demand for that sector. Since the demand categories are themselves assumed to be "independent," the specific-electric elasticities can be computed as the quantity-weighted average of the sector elasticities, with the quantities being the amount of "specific electricity" demanded in the corresponding sector for some base year. Using 1972 as a base year and assuming that in that year in the United States all electric consumption except one-half the residential/commercial space heating, water heating, and cooking electricity consumption in that year was "specific-electric" consumption, the elasticities for the "specific-electric" category shown in table 2.3 were calculated.

Table 2.3.
Demand Functions for Per Capita Net Energy Used in Programming Model

Sector	Price elasticity	Per capita income elasticity	Population elasticity
Specific electric	−0.65	0.92	1.00
Industry	−0.52	0.76	1.00
Residential and commercial	−0.79	1.08	1.00
Transportation	−1.28	0.81	1.00

DERIVATION OF ELASTICITIES: Elasticities are derived from those presented in table 2.2. The conceptual basis is different, however, in that the elasticities in table 2.2 include electricity, both specific and general, in the sectoral energy consumption data. To obtain the four different elasticities from the three estimates required making an assumption about the relative elasticities. See text.

3 AVAILABILITY OF ENERGY RESOURCES AND ALTERNATIVE ENERGY SUPPLY TECHNOLOGIES

The central problem in the efficient use of energy resources is that those resources that require little labor and capital to recover and use are, ultimately, severely limited in supply. And those resources that are superabundant have technologies which require substantial research and development and are likely to need much higher capital costs than the fuels they replace. This chapter reviews the principles and data used in this study to estimate the availability and costs of recovering alternative resources and transforming them into useful energy. Part A discusses resource availability, while part B considers alternative energy supply technologies.

A. *Availability and Cost of Energy Resources*

One theme of this study is that energy resources are limited in supply and that those resources that are most economical today are also those that have the most severe ultimate limitations on availability. In the long run, the question of availability rests on two separate issues: First, what is the natural availability of different resources at different grades or concentrations? Second, at what price can these different grades be extracted and transformed into economically useful energy resources?

The central difficulty in making resource estimates is that

they rest on unverified theories or extrapolations of past experience. Thus some estimates of the natural availability of different resources are often "pessimistically" limited to those resources that have actually been discovered or proved; while others "optimistically" interpolate between current proved reserves and crustal abundance by assuming a constant grade-availability relationship. Either projection rests on little firm knowledge about the exact resources being presupposed. Similarly, estimates of the future costs of extraction and beneficiation of low grade ores rest on shaky ground as little or no experience has been gained at these very low grades. For example, the currently mined grades of uranium ores are around 1,500 parts per million (ppm), while future resources are contemplated to a grade level of as low as 5 ppm, without solid knowledge about the actual processes and their environmental impacts.

A.1. *Analytical Issues*

It is useful to begin with a brief review of general principles of mineral recovery. The starting point of a discussion of costs is well summarized by Fettweis:

> The useable minerals whether they be copperbearing minerals or coal are known to be unevenly distributed within the earth's crust. This applies to the concentration per unit volume, the grade, in the same way as to the volumetric size of the geological bodies with a certain average content, and, therefore, also to the quantities. Both kinds of distribution seem to be log-normal, which also means that of a large population of deposits a few contain most of the material. In general, as far as the economic useability of mineral deposits is concerned, it can be claimed that the grade of concentration and the tonnage is of special importance, as well as the grade-tonnage ratio.[1]

With this general view in mind, the *availability function* for a resource is defined as: the functional relationship between the cumulative extraction of a given resource at a point of time and

---

1. From Fettweis (1975) with minor changes. Also see McKelvey (1972), Ahrens (1954), and Brinck (1967).

determining variables. The determining variables include: (1) the natural availability of the resource; (2) the distribution of the size and grade of deposits of the resource; (3) the state of the technology for finding and extracting the resource; (4) the marginal cost of the resource; and (5) the probability that the cumulative extraction will be as high as the given amount. Thus (1) the natural availability determines the overall amount that can be extracted; while (2) the distribution of the total determines the amount that will be concentrated in high-grade (low-cost) deposits; (3) the state of technology determines the ease with which new deposits can be uncovered, as well as the cost of mining and beneficiating the ores; (4) the marginal cost determines how far producers will be willing to go down the grade curve; and finally, (5) the probability is a reflection of the uncertainty that exists about the cumulative extraction at each level of the other four independent variables, given the other factors in (1) through (4).

A.2. *A Simple Model of the Cost of Extracting Low-grade Reserves*

Despite the enormous uncertainties about the availability function for energy resources, in order to proceed it is necessary to make a judgment about the availability of energy resources. Therefore, I have constructed a very simple model of future costs. The model rests on geochemical principles as well as observed regularities, but it must be emphasized that it is nevertheless extremely speculative. The fundamental assumptions are:

1. It is assumed that the concentration of each resource is lognormally distributed around the mean concentration of the resource, and that the size of deposit in which a uniform concentration exists is lognormally distributed around the mean deposit size.

2. It is assumed that the extraction and beneficiation (upgrading) cost of a given resource grade is inversely related to the concentration of the resource.[2]

---

2. The assumptions are based on work of McKelvey (1960), Coulomb (1959), Allais (1957), Brinck (1967), and Grenon (1975).

In applying these assumptions, the techniques will be slightly different for resources such as oil and gas, where there is very little variance in the grade of the resource but where the major dispersion comes in the size of the deposit; and for mineral resources (shale, coal, uranium, thorium), where the major difference arises in the grade rather than in the size of the deposit. In the former case, the difficulty lies in the cost of finding and recovering small or frontier deposits, while in the latter, the difficulty lies in extracting and beneficiating low-grade ores.

First, in discussing the principles for recovery of oil and gas, I will rely heavily on the theoretical work of Gordon Kaufman and empirical findings of M. K. Hubbert.[3] The general proposition of Kaufman is that, although pools are lognormally distributed in a given geological environment, the probability of finding a pool is proportional to its size and that this sequential sampling occurs without replacement. Thus the large pools tend to be discovered first, and the cost of discovering pools increases as the discovery process occurs. Therefore, even though the size of deposits is lognormally distributed, the time-path of discovery produces a distribution that is the convolution of a lognormal distribution and the proportional probability of discovery. Kaufman has estimated the distribution of discoveries both analytically and by Monte Carlo methods. The striking feature of his results is the sharp drop in the finding rate for new pools after the first few finds.

In what follows, I simplify Kaufman's work by adopting the negative exponential distribution as a reasonably accurate approximation of the theoretical distribution of the finding rate. In work going back to the mid-1950s, Hubbert presented evidence that the discoveries of petroleum per foot drilled have followed a negative exponential function. Thus, let h be cumulative footage drilled, $Q(h)$ be cumulative discoveries, and $Q'(h) = dQ/dh$ be oil discovered per foot drilled. If h_0 feet have been drilled at a given time, the marginal discoveries

3. Barouch and Kaufman (1976), see Hubbert (1969), Menard and Sharman (1975).

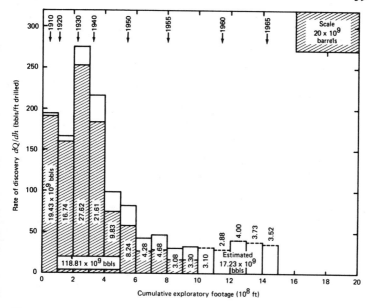

Figure 3.1.
Crude oil discoveries per foot of exploratory drilling footage in the United States, exclusive of Alaska, 1860–1967. From Hubbert (1969).

per foot drilled, $Q'(h_0)$, according to Hubbert, follow the relation:[4]

$$Q'(h_0) = \alpha e^{-\beta h_0}. \tag{3.1}$$

Figure 3.1 shows the results of Hubbert's investigation, indicating very clearly the decline in the finding rate as drilling progresses.

A similar hypothesis has been tested more recently by the National Petroleum Council in its studies of future oil availability for separate regions of the United States.[5] These studies investigate the finding rate of oil in place as a function of time

4. Hubbert (1969).
5. National Petroleum Council (1972).

(rather than cumulative footage drilled). Figure 3.2 shows the finding rate for the New Mexico–Texas region as shown in NPC (1972). Although these data do not correspond precisely to the hypothesis in equation (3.1), evidence of a declining relation is clear.

Although Hubbert's work is quite interesting, it is deficient in several respects. First, as the NPC study indicates, discoveries are in terms of oil-in-place rather than produced oil. Thus a recovery factor should be included. Second, the Hubbert curve in figure 3.1 confounds two trends—the depletion effect and improvements in technology. Although depletion occurs with increased drilling—thus leading to lower finding rates—improvements in geological knowledge and drilling techniques would shift upward the Hubbert curve in figure 3.1. Further, it is likely that at a given point in time the short-run finding function is steeper than the Hubbert curve, but improvements in drilling technology continuously shift the curve up over time.

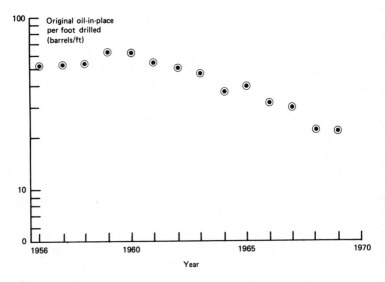

Figure 3.2.
Finding rate for oil-in-place as a function of time, NPC region 5 (Southeast Texas, West New Mexico), from NPC (1972).

Third, to obtain the resource costs of petroleum, we must translate footage per barrel into costs per barrel; again, improvements in the technology of drilling will generally lower the resource costs per foot drilled.

Given that we accept the Hubbert hypothesis in equation (3.1) about the relation between discoveries of oil in place and footage, we can calculate the ultimate recoverable reserves as follows. Let $Q(h)$ be cumulative oil-in-place discovered as a function of footage drilled. We then have:

$$\begin{pmatrix} Ultimate \text{ discoverable} \\ \text{oil-in-place} \end{pmatrix} = Q(\infty) = \int_0^\infty \alpha e^{-\beta h} dh = \frac{\alpha}{\beta},$$

and recoverable oil-in-place is

$$R(\infty) = kQ(\infty),$$

where k is the recovery rate. In a more complete model, we should take account of the efficiency of recovery techniques. Improvement in exploration has two functions. First it increases the fraction of the total oil-in-place that is discoverable. [Hendricks (1965) for example, assumes that only 60 percent of all oil-in-place is discoverable.] Second, it decreases the amount of footage required to discover oil-in-place. In practice, it is difficult to disentangle these effects, so in what follows it is assumed that the cost of drilling may decrease, but that the finding function in equation (3.1) does not shift over time.

Finally, we can derive the *cost function* as follows: Let $MC[R(h)]$ be the marginal cost of an additional unit of recoverable oil-in-place discovered. Then if $\bar{h}$ units have been drilled, the marginal cost is

$$MC[R(\bar{h})] = \frac{c(t)}{\alpha} \frac{R(\infty)}{R(\infty) - R(\bar{h})},$$

where $c(t)$ is the cost per foot drilled. (Clearly, in the complete calculation, $c(t)$ includes the exploration, development, and production costs, properly discounted.) Or if $R[h(0)]$ units have been recovered to date, the ratio of future marginal cost $\{MC[R(h(t))]\}$ at time t to current marginal cost $\{MC[R(h(0))]\}$ is:

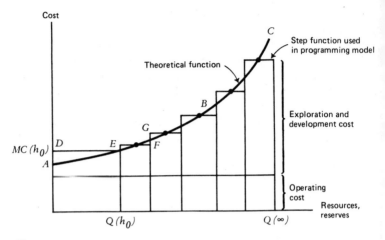

Figure 3.3.
Illustration of technique used to approximate the nonlinear cost function for use in the mathematical programming model.

$$\frac{MC[R(h(t))]}{MC[R(h(0))]} = \frac{c(t)}{c(0)} \cdot \frac{R(\infty) - R(h(0))}{R(\infty) - R(h(t))} \tag{3.2}$$

This particularly simple formula relies only on the ratio of remaining reserves and on the cost of drilling.

The cost function is shown in figure 3.3. In general, the function will have the nonlinear shape shown as the function ABC. In the calculations, however, this is approximated by the step function, DEFG.... Further note that, in general, extraction has three stages: exploration, development, and production. I have assumed that both exploration and development follow the negative exponential relationship, while operating costs are constant.[6]

To estimate costs, go through the following steps. For each country and each resource the amount of ultimately producible but as yet undrilled resource is estimated, as well as the cumulative production. In addition, the current marginal cost of the resource is estimated. Then break the cost function

6. See particularly Adelman (1972).

into a small number of steps and evaluate the cost of extraction by using equation (3.2), as well as an assumed rate of decrease of the cost of drilling over time. The approach just described applies to oil and gas, but the same general approach has been used for mineral resources as well.

The issue of uncertainty is more intractable. In principle, all the parameters of the relations (cost, discoverable resources, etc.) are imperfectly known, sometimes with considerable uncertainty. As for other variables, I use the expected value of all variables in the basic estimates presented in chapters 5 through 8. Explicit treatment of uncertainty is presented in MRG (1977) and Nordhaus and van der Heyden (1977).

A.3. *Resource Estimates*

Tables 3.1 and 3.2 present the details of the cost and resource estimates for individual countries and for different resources as used in this study. The estimates are very crude and approximate, both because I am not an expert on resource availabilities and because many of the required figures are unavailable. For example, while estimates for uranium are carefully compiled for the United States, there are no such careful studies available for thorium or for uranium in other countries.

B. *Alternative Energy Supply Technologies*

This section explains the data used for describing energy technologies. Extensive if somewhat conflicting data are available on most of the important processes. For the purposes of this study, these are separated into (1) labor and other current costs, (2) capital costs, and (3) fuel inputs. All processes are in terms of delivered energy (such as kilowatt-hours of electricity). Capital costs are considered to be invested at the beginning of the period, but explicit accounting for the capital stock—as in a vintage model—is not performed. Rather, it is assumed that there are no constraints on the capital goods industry. Labor and other current costs are spent continuously; and explicit accounting is made for thermal efficiency.

The uncertainties of the cost estimates in tables 3.3 through 3.8 vary among the different technologies. In principle, they

Table 3.1.
*Cost of Intermediate Energy Products Exclusive of Royalties, by Source**

Energy source	Cost in 1975 dollars (per million btu)	Cost in 1975 dollars (per conventional unit)[†]
CRUDE PETROLEUM delivered[‡]		
United States[§]		
Category 1 drilled reserves	0.20	1.14
Category 2 undrilled reserves	1.51	8.78
Category 3 undrilled reserves	1.77	10.26
Category 4 undrilled reserves	2.14	12.41
Category 8 undrilled reserves	12.90	74.90
Persian Gulf and North Africa		
Category 1 drilled reserves	0.01	0.07
Category 2 undrilled reserves	0.11	0.64
CRUDE OIL FROM SHALE[‖]		
United States, delivered		
30 gallons per ton of shale	1.90	11.05
15 gallons per ton of shale	2.77	16.05
COAL		
United States, delivered		
Liquefied[‖]	2.72	15.80
Gasified[‖]	2.77	2.77
Strip-mined	0.81	18.20
Deep mined	1.25	28.10
BASELOAD ELECTRICITY GENERATION		
Petroleum (fuel at $2/mmbtu)	6.90	23.6
Coal (fuel at 80¢/mmbtu)	4.94	16.9
Light water reactor	5.17	17.7
Breeder reactor[‖]	5.46	18.7
HYDROGEN (BY ELECTROLYSIS)[‖]	8.70	. . .

SOURCES: Various published and unpublished sources. Details available in appendix available from author on request.

*Electricity costs are busbar. Costs include direct costs and current costs, but exclude any shadow prices or royalties. The real cost of capital is taken to be 13%.

[†] Per barrel for crude petroleum, crude oil from shale, and liquefied coal; per thousand cubic feet for gasified coal; per ton for stripmined and deepmined coal; per 1,000 kilowatt-hour for electricity generation.
(continued)

Table 3.2.
Energy Resources, by Region and Resource, 1970

	U.S.	OPEC	ROW	Total
Petroleum (10^{15} btu)				
Proven	220	1,978	805	3,003
Unproven	876	2,134	5,701	8,711
Shale (10^{15} btu)				
High grade	853	0	4,321	5,174
Low grade	14,256	0	79,814	94,070
Coal (10^{15} btu)				
Surface	6,100	0	10,625	16,725
Deep	13,250	0	23,075	36,325
Natural gas (10^{15} btu)				
Proven	265	870	1,013	2,148
Unproven	669	2,975	5,070	8,714
Natural uranium (10^6 metric tonnes)				
Grade 1 ($0–25/lb)	2.8	0	4.5	7.3
Grade 2 ($25–50/lb)	1.2	0	2.3	3.5
Grade 3 ($50–100/lb)	2.3	0	4.4	6.7
Grade 4 ($100–200/lb)	8.1	0	13.6	21.7
Grade 5 ($200–500/lb)	1,700	0	3,400	5,100
Thorium (10^6 metric tonnes)				
Grade 1 ($0–25/lb)	0.6	0	1.5	2.1
Grade 2 ($25–50/lb)	0.8	0	1.6	2.4
Grade 3 ($50–100/lb)	10.8	0	20.6	31.4
Grade 4 ($100–200/lb)	1,100	0	1,890	2,990
Grade 5 ($200–500/lb)	5,600	0	10,800	16,400

SOURCES: From various published and unpublished sources. Appendix available from author on request.

ROW: rest of world.

[‡] Petroleum was separated into eight different cost categories for the United States and two categories for all other regions. Category 1, the lowest cost, is drilled reserves, while categories 2 through 8, the highest cost, are undrilled.

[§] Costs for petroleum annualized at 6% discount rate, 9.5% decline rate, and 3% annual rise in real oil price.

[‖] Technologies are immature. Estimates refer to the expected value of costs after large scale commercialization has taken place..

are designed to represent the cost of technologies with *current* knowledge, even though the engineering details have not been worked out and therefore some development costs must be incurred before full-scale production can take place. On the other hand, technologies have been graded by the degree of development. The degree of development ranges from A, which is for large-scale production today, such as the Otto cycle engine or coal-fired electricity generation plant, to E, in which the technology is speculative and the technical feasibility has not been proved, this being the case for fusion power. These classifications can be thought of as reflecting the uncertainty in the cost of production of different processes or as uncertainty about the technical coefficients.

For processes where alternative estimates are available, I have estimated the "unreliability factor," which is generally the standard deviation of the logarithm of estimated cost from different sources.

A basic purpose of this study is to identify the existence of backstop technologies. In this, it is possible to work backward from the demand categories rather than through the technologies to the resource base. For the first three demand categories (electric, industrial, and residential), according to the classification used here, there will exist backstop technologies if there is a backstop technology for electricity generation. According to the classification used above, there are no class A backstop technologies for electricity generation; there are three class B technologies, all of them nuclear (LWR-U plus low-grade uranium is rated B, as are the different breeders); there is in addition one backstop C class (solar electricity), and one E class (fusion power).

The remaining category is demand category four—transportation. For transportation, three possible backstop technologies have been identified. (1) Hydrogen fuel can be substituted for all uses, but its development is speculative. While there are reasonably firm estimates for the production costs, the problems of distribution and storage in small units are large. Thus transportation using hydrogen is based on a class B backstop technology (electricity generation) plus a class C utilization technology (hydrogen automobiles); thus overall it rates a C. (2) Methanol—a good substitute for gasoline—can be synthesized *if* an inexhaustible source of carbon can be found. At present, there are no estimates of the cost of extracting carbon from the sea, from rocks, or the atmosphere. As long as run costs of methanol are speculative, it rates only a D. (3) An all-electric automobile can be used. These vehicles are relatively inexpensive, but at present the specific-power and the specific-energy of their batteries are well below that of internal combustion engines; performance is therefore quite dismal. An adequate electric car, therefore, rates no better than a D. Overall, then, the backstop technology for transportation rates at best a C.

The data for the alternative processes are given in tables 3.3 through 3.8. Transportion costs were calculated from actual as well as engineering data. The resources and demand for each region were considered to be at a point (thus U.S. demand was assumed to be 200 overland miles from New York, while U.S. petroleum was assumed to lie 200 overland miles from the Gulf Coast). Using the calculated distances and engineering estimates of costs, the cost per unit was calculated according to table 3.7. Distribution costs (inclusive of taxes) are shown in table 3.8.

Table 3.3.
Technologies for Electricity Generation (All prices in 1975 U.S. dollars)

	(1) Current costs (dollars per 1,000 kwh)	(2) Capital costs (dollars per kw–capacity)	(3) Lifetime (years)	(4) Thermal efficiency (fraction)	(5) Operating rate (fraction)	(6) Year of availability	(7) Technical classification
Fossil							
Coal with sulfur removal							
Value	0.7	450	30	0.40	0.75	current	A
Unreliability factor		(0.06)					
Oil							
Value	0.7	280	30	0.40	0.75	current	A
Unreliability factor		(0.13)					
Gas							
Value	0.7	275	30	0.40	0.75	current	A
Unreliability factor		(0.03)					

Nuclear fission

						current	A
LWR	2.1	545	30	n.a.	0.75	1980	B
Unreliability factor		(0.03)					
LWR-Pu	2.1	530	30	n.a.	0.75	1995	B
HTGR-35	1.4	545	30	n.a.	0.75	1995	B
HTGR-33	1.1	570	30	n.a.	0.75	1995	B
LMFBR-Pu	1.4	730	30	n.a.	0.75	2015	C
Unreliability factor		(0.14)					
LMFBR-adv	1.2	800	30	n.a.	0.75	1995	B
HWR	1.8	735	30	n.a.	0.75	2005	C
HTBR	1.0	775	30	n.a.	0.75	2000	D
Best of solar or fusion	4.0	1,730	20	0.15	0.75		

SOURCES: Sources for tables 3.3 through 3.8 available on request from the author.

KEY: LWR = light water reactor; LMFBR-Pu = early version of liquid metal fast breeder reactor, plutonium fueled; HTGR-35 = high-temperature gas-cooled reactor, U-235 fueled; HTBR = high-temperature gas-cooled breeder reactor; LWP-Pu = light water reactor, plutonium recycle; HTGR-33 = high-temperature gas-cooled reactor, U-233 fueled; HWR = heavy water reactor; LMFBR-adv = advanced liquid metal fast breeder reactor.

n.a. = not applicable

Table 3.4.
Technologies for Conversion, Nonelectric Processes (All prices in 1975 U.S. dollars)

Transformation activities	Current costs, except fuel (dollars per 10^6 btu)	Capital (dollars per 10^6 btu/yr)	Depreciation rate (per yr)	Thermal efficiency (fraction)	Operating rate (fraction)	Year of availability*	Technical classification†
1. Coal gasification							
Value	0.17	5.52	0.05	0.62	0.9	1987	C
Unreliability factor‡	(0.97)	(0.42)	—	(0.05)	—		
2. Coal liquefaction							
Value	0.39	6.34	0.05	0.67	0.9	1995	C
Unreliability factor	(0.36)	(0.23)	—	(0.04)	—		
3. Petroleum refining	0.025	1.48	0.05	0.9	0.9	current	A
4. Hydrogen by electrolysis	0.36	2.33	0.05	0.68	0.9	current	A
5. Thermal hydrogen by HTGR	0.23	0	0.05	1.25	0.9	2030	E

* Availability for new technologies defined as that point in time when the production capacity reaches 1.0 quads per year. This corresponds to approximately sixteen 1,000 MW^e plants or approximately five 100,000 barrel/day synthetic petroleum plants.

† Technical classification is as follows: A = in large scale use; B = in pilot plant; C = in laboratory; D = speculative, but technologically unproven; E = technologically unproven.

‡ Unreliability factor is the standard deviation of the logarithm of estimated costs from alternative sources.

HTGR = high temperature gas-cooled reactor.

Table 3.5.
Materials Balance for Nuclear Fuel Cycle (Inputs and outputs, thousands of metric tons per 10^{15} btu/yr or per 10^{15} btu/yr capacity)

Reactor		Natural uranium	Thorium	Plutonium	U-233
LWR	Initial	−17.07	—	—	—
	Annual	−6.23	—	0.0075	—
	Final	10.80	—	0.0075	—
LMFBR-Pu	Initial	—	—	−0.0835	—
	Annual	—	—	0.00294	—
	Final	—	—	0.086	—
HTGR-35	Initial	−13.61	−1.037	—	—
	Annual	−1.81	−0.023	—	—
	Final	11.79	0.949	—	—
HTBR	Initial	—	−2.	—	−0.15
	Annual	—	−0.02	—	—
	Final	—	—	—	0.15
LWR-Pu	Initial	−8.07	—	−0.071	—
	Annual	−2.95	—	−0.0075	—
	Final	5.17	—	0.063	—
HTGR-33	Initial	−13.61	−1.037	—	—
	Annual	−2.82	−0.023	—	0.0096
	Final	10.89	0.949	—	0.01
HWR	Initial	−21.04	−1.	—	—
	Annual	−8.09	−0.02	—	—
	Final	20.23	0.98	—	—
LMFBR-adv	Initial	—	—	−0.0628	—
	Annual	—	—	0.00432	—
	Final	—	—	0.064	—

NOTES: Figures in table give inputs as negative terms, outputs as positive items. "Initial" denotes the required initial inventory at the beginning of the 30-year operation in 10^3 tonnes per 10^{15} btu/yr. capacity. "Annual" denotes the annual feed required per year per kilowatt-electric-year over the 30-year lifetime of the plant in 10^3 tonnes per 10^{15} btu. "Final" indicates the final inventory to be withdrawn at the end of the 30-year operation in 10^3 tonnes per 10^{15} btu/yr. capacity.

KEY: LWR = light water reactor; LMFBR-Pu = early version of liquid metal fast breeder reactor, plutonium fueled; HTGR-35 = high-temperature gas-cooled reactor, U-235 fueled; HTBR = high-temperature gas-cooled breeder reactor; LWP-Pu = light water reactor, plutonium recycle; HTGR-33 = high-temperature gas-cooled reactor, U-233 fueled; HWR = heavy water reactor; LMFBR-adv = advanced liquid metal fast breeder reactor.

Table 3.6.
Technologies for Utilization (Costs in 1975 U.S. dollars)

	Incremental current (per million btu)	Incremental capital (per million btu/yr capacity)	Lifetime (years)	Thermal efficiency (fraction)	Operating rate (fraction)	Year of availability	Technical classification
Industrial							
Sulfur abatement for (per million btu thermal)							
Coal	0.	1.33	30.	1.0	0.75	current	B
Petroleum	0.	0.44	30.	1.0	0.75	current	B
Transportation							
Electric automobile*							
Present	7.72	47.06	5.	0.59	1.0	current	B
Future	7.72	37.65	5.	0.59	1.0	2000	D
Hydrogen automobile	2.82	23.53	5.	0.21	1.0	2000	C
Gasoline-methanol automobile							
Otto cycle	0.00	0.00	5.	0.19	1.0	current	A
Diesel	−6.68	11.30	5.	0.23	1.0	current	A

*Substandard performance.

Table 3.7.
Transportation Costs

	Cost *(1975 U.S. mils per 100 miles* *per 10^6 btu, except nuclear)*
Petroleum	
Overland pipeline	5.0
Tanker ship (less than 2,000 miles)	3.0
(more than 2,000 miles)	1.5
Coal	
Unit train	30.0
Collier	3.0
Pipeline	25.0
Natural gas	
Overland pipeline	20.0
LNG Tanker	15.0
Nuclear fuel (per gram fissile)	3.0

Table 3.8.
Taxes and Distribution Costs for Demand Functions (1975 U.S.
dollars per 10^6 btu input)

	Demand sector			
	Specific *electric* *(1)*	*Industry* *(2)*	*Residential* *and* *commercial* *(3)*	*Transportation* *(4)*
Local distribution *costs and taxes*				
Petroleum	n.a.	0.07	0.83	2.93
Gas	n.a.	0.17	0.97	0.97
Coal	n.a.	0.39	0.66	0.39
Electricity	3.32	1.38	4.49	4.49
Hydrogen	n.a.	0.22	1.45	1.48

4 DETAILED EQUATIONS OF THE ENERGY MODEL

The two previous chapters outlined the principles and data underlying two important components of the energy model— the demand functions and the technological relations. This chapter outlines in detail the exact equations used.[1]

This chapter is basically a technical description of the model for those who require a detailed view of the exact procedures or methodology.[2] Sections A to E list the equations used in the model and can be omitted without substantially losing the thrust of the argument. Section F provides a simple example of the methodology with interpretation and would be worth some study by those who would like to understand exactly how linear programming can provide estimates of the allocation and pricing of exhaustible resources.

A. *Indices Used in Equations*

I start with the indices used for the variables:

A.1. *Country of Origin of Energy Resources*

$i = 1, 2, 3$: the United States (U.S.), OPEC, the rest of the world (ROW).

---

1. The computer code used to generate the linear programming code is available on request in Nordhaus (1978).

2. Ludo van der Heyden graciously assisted in preparation of the exposition of the detailed equations of the model.

A.2. *Resource Types*

$j = 1, 2, 3, 4$: oil, shale, coal, gas (fossil resources).

$j = 5, 6$: natural uranium, thorium (natural nuclear resources).

$j = 7, 8$: plutonium, uranium 233 (produced nuclear resources).

A.3. *Grades*

$jj = 1, 2$: high, low (for coal and shale resources).

$jj = 1, \ldots, 5$: highest, ..., lowest (for nuclear resources).

$jj = 1, \ldots, 10$: highest, ..., lowest (crude oil and natural gas—varies with exact specification).

A.4. *Reactor Types*

$jr = 1, \ldots, 8$: 2 light water reactors; 2 liquid metal fast breeder reactors (plutonium); 3 high temperature gas reactors; heavy water reactor (see chapter 3).

A.5. *Fuels*

$k = 1, \ldots, 5$: oil, gas, coal, electricity, hydrogen.

A.6. *Destinations Where Energy Is Consumed*

$\ell = 1, 2$: U.S., ROW.

A.7. *Demand Categories*

$m = 1, \ldots, 4$: specific electricity, industrial heat, residential heat, transportation (see chapter 2).

A.8. *Demand Interpolation Steps*

$mm = 1, 2, \ldots, 24$: number of stepwise approximations to the demand function.

A.9. *Time Periods*

$n = 1, 2, \ldots, 12$: 1970–79, 1980–89, ..., 2080–89.

$nr = 1, \ldots, nrf$: number of periods after oil and gas drilling occurs, to use in calculating decline curve ($nrf = 5$).

A.10. *Backstop Technologies*
  $jb = 1, 2$: solar, fusion depending on experiment.

B. *Activities*
B.1. *Extraction and Transportation of Natural Resources, Conversion of Resources into Fuels*
  $x(i,j,jj,k,\ell,n)$ = extraction of (fossil) resource j, grade jj, in area i, transportation to area ℓ, for conversion into fuel k, all activities taking place in time period n ($j = 1, \ldots, 4$ only), 10^{15} btu.

  $m(i,j,jj,\ell,n)$ = mining of (nuclear) resource j, grade jj, in area i, transportation to area ℓ, for conversion into electricity (fuel 4), all activities taking place in time period n ($j = 5, 6$ only), thousands of metric tonnes.

  $ntr(i,j,\ell,n)$ = transportation of (nuclear) resource j from area i to area ℓ in time period n ($j = 5, \ldots, 8$), thousands of metric tonnes.

  $ne(jr,\ell,n)$ = nuclear electricity generation in area ℓ and time period n, using reactor type jr, 10^{15} btu electric.

  $hyd(\ell,n)$ = hydrogen production in area ℓ and time period n from either electricity or heat, 10^{15} btu.

  $bck(jb,k,\ell,n)$ = production in area ℓ and time period n of fuel k using backstop technology jb, 10^{15} btu.

B.2. *Processing of Fuels into Demand Categories*
  $xp(k,\ell,m,n)$ = processing of fuel k to meet demand category m in area ℓ and time period n, 10^{15} btu.

B.3. *Energy Consumption*
  $xc(\ell,m,mm,n)$ = consumption in region ℓ, for demand category m, demand interpolation step mm, and time period n, 10^{15} btu of net energy (see chapter 2).

C. *Constraints*
  The most important part of the model is the set of re-

source, conversion, and delivery constraints. This is the set of activities and constraints that takes resources in the ground, converts them to useful fuels, and delivers them to a given demand sector. It is represented in sections C.1 through C.3.

C.1. *Resource Constraints*
C.1.a. Fossil resources

$$\sum_{k,\ell,n} x(i,j,jj,k,\ell,n) \leqq R(i,j,jj) \qquad\qquad (j = 1, \ldots, 4)$$

C.1.b. Nuclear resources

$$\sum_{\ell,n} m(i,j,jj,\ell,n) \leqq R(i,j,jj) \qquad\qquad (j = 5, 6),$$

where

$R(i,j,jj) = $ total availability of resource j, grade jj, in area i.

Constraint C.1.a keeps track of the (exogenously given) resource endowments for each resource, grade, and region (see chapter 3 for details). Thus

$$\sum_{k,\ell,n} x(1,1,1,k,\ell,n) \leqq R(1,1,1)$$

is the equation that ensures that total extraction of low-cost U.S. petroleum—the left-hand side of the inequality—does not exceed total availability, $R(1,1,1)$. Separate totals are kept for each resource.

The interpretation of the nuclear availability is as follows: constraint C.1.b is the resource availability constraint for natural uranium and thorium. It is exactly analogous to that for the fossil fuels, but enters separately.

C.2. *Conversion Constraints*
C.2.a. Oil, gas, coal

$$\sum_{m} xp(k,\ell,m,n) - \sum_{jb} bck(jb,k,\ell,n)$$

$$- \sum_{j=2,3} \sum_{i,jj,n} feff(j,k) \cdot x(i,j,jj,k,\ell,n)$$

$$- \sum_{j=1,4} \sum_{i,jj} \sum_{nr=1}^{nrf} rf(i,j,nr) \cdot feff(j,k) \cdot x(i,j,jj,k,\ell,$$

$$n + 1 - nr) \leqq 0 \qquad\qquad (k = 1, 2, 3).$$

Note: All variables with a negative or zero time period index are equal to 0 (i.e., are omitted).

C.2.b. Electricity

$$\sum_m xp(k,\ell,m,n) + hyd(\ell,n) - \sum_{jb} bck(jb,k,\ell,n)$$

$$- \sum_{jr} ne(jr,\ell,n) - \sum_{i,jj} feff(3,k) \cdot x(i,3,jj,k,\ell,n) \leqq 0$$

$$\text{(electricity, } k = 4).$$

C.2.c. Hydrogen

$$xp(k,\ell,4,n) - \frac{1}{heff} hyd(\ell,n) - \sum_{jb} sol(jb,k,\ell,n) \leqq 0$$

$$\text{(hydrogen, } k = 5),$$

where

$feff(j,k) =$ efficiency factor for converting resource j into fuel k,

$heff =$ efficiency factor for converting electricity into hydrogen, fraction.

$nrf =$ number of time periods for the recovery from oil and gas well (see section A.9), years.

$rf(i,j,nr) =$ recovery factor in country i for resource j, $(nr - 1)$ time periods after start of exploitation $(j = 1, 4)$, fraction.

Constraints C.2.a, C.2.b, and C.2.c made sure that the total availability of a particular fuel in a region at a given time is at least as great as the deliveries. Thus if only low cost U.S. petroleum was used as oil for U.S. transport period 1, the equation would read:

$$xp(1,1,4,1) - feff(1,4) \cdot x(1,1,1,1,1,1) \leqq 0.$$

As long as the resource is scarce, the equality will hold. Note that constraints C.2.b and C.2.c—for electricity and hydrogen —keep track of nuclear generation of electricity and conversion to hydrogen.

C.3. Delivery Constraints

$$\sum_{mm} xc(\ell,m,mm,n) - \sum_{k} \frac{1}{peff(k,m)} xp(k,\ell,m,n) \leq 0,$$

where

$peff(k,m)$ = thermal efficiency for flow of fuel k to demand category m, fraction.

Constraint C.3 keeps the accounts for deliveries to consumption. The second term is the total deliveries to demand sectors from all fuels. The second term must be no less than the first term—total consumption in a given demand category, region, and time.

C.4. Import Constraints (U.S. Only)

$$\sum_{j=1}^{4} \sum_{jj,k,n} x(2,j,jj,k,1,n) \leq IMPT(n),$$

where

$IMPT(n)$ = upper limit on U.S. imports for fossil resources in time period n, 10^{15} btu.

Constraint C.4 allows different policies to be introduced for restricting imports or estimating limit prices on oil imports.

C.5. Constraints on U.S. Energy Consumption

$$\sum_{jr} nef \cdot ne(jr,\ell,n) + \sum_{nr=1}^{nrf} \sum_{j=1,4} \sum_{i,jj,k} rf(i,j,nr) \, x(i,j,jj,k,\ell,$$

$$n+1-nr) + \sum_{j=2}^{3} \sum_{i,jj,k} x(i,j,jj,k,\ell,n) \leq BTU(n),$$

where

$BTU(n)$ = upper limit on U.S. fossil fuel and electricity production in time period n (not including backstop technologies—solar and fusion), 10^{15} btu/yr.

nef = conversion factor for nuclear electricity generation (10,000 btu/kwh), btu per btu-electric.

Constraint C.5 allows experiments to be imposed that calculate the loss in real income originating from alternative energy growth constraints.

C.6. *Constraints on Growth of Production (Extraction Technologies)*

$$- \sum_{jj,\ell} g \cdot m(i,j,jj,\ell,n-1) + \sum_{jj,\ell} m(i,j,jj,\ell,n) \leq GP(i,j,n)$$

$$(j = 5, 6; n = 1, 2, \ldots),$$

and

$$- \sum_{jj,k,\ell} g \cdot x(i,j,jj,k,\ell,n-1) + \sum_{jj,k,\ell} x(i,j,jj,k,\ell,n)$$

$$\leq GP(i,j,n) \qquad (j = 1, \ldots, 4; n = 1, 2, \ldots),$$

where

$GP(i,j,n)$ = production level at which the technology associated with the extraction of resource j in area i and time period n becomes available, 10^{15} btu per period.

g = maximal production growth factor for extraction technologies, growth rate per period.

Note: If the technology associated with the exploitation of resource j in area i becomes available in time period $nav(i,j)$ at a production level $PRODAV(i,j)$, then the coefficients $GP(i, j,n)$ are:

$$GP(i,j,n) = 0 \qquad\qquad , n < nav(i,j),$$
$$= PRODAV(i,j), n = nav(i,j),$$
$$= STEP(i,j) \qquad , n > nav(i,j),$$

where $STEP(i,j)$ is a small but positive number, which allows an efficient allocation of resources to start the exploitation of resource j after time period $nav(i,j)$, if it is efficient.

The role of the growth constraints (C.6) is to ensure that production and new technologies do not grow at an unrealistic rate. These become particularly critical for evaluation of new technologies [see MRG (1977)].

C.7. Constraints on Growth of Production (New Technologies)

C.7.a. Coal gasification

$$- \sum_{i,jj} gn \cdot x(i,3,jj,2,\ell,n-1) + \sum_{i,jj} x(i,3,jj,2,\ell,n)$$
$$\leq GS(\ell,n),$$

where

$GS(\ell,n)$ = production level at which the coal gasification technology becomes available in area ℓ and time period n, 10^{15} btu/period.

gn = production growth factor for new technologies, growth rate per period.

C.7.b. Coal liquefaction

$$- \sum_{i,jj} gn \cdot x(i,3,jj,1,\ell,n-1) + \sum_{i,jj} x(i,3,jj,1,\ell,n)$$
$$\leq Lq(\ell,n),$$

where

$Lq(\ell,n)$ = production level at which the coal liquification technology becomes available in area ℓ and time period n, 10^{15} btu per period.

C.7.c. Nuclear electricity generation

$$-gn \cdot ne(jr,\ell,n-1) + ne(jr,\ell,n) \leq NEG(jr,\ell,n),$$

where

$NEG(jr,\ell,n)$ = production level at which the nuclear reactor type jr becomes available in area ℓ and time period n, 10^{15} btu electric per period.

C.7.d. Backstop technologies

$$-\sum_k gn \cdot bck(jb,k,\ell,n-1) + \sum_k bck(jb,k,\ell,n)$$
$$\leq BACK(jb,\ell,n),$$

where

$BACK(jb,\ell,n)$ = production level at which the backstop technology jb becomes available in area ℓ and time period n, 10^{15} btu per period.

The coefficients $GS(\ell,n)$, $Lq(\ell,n)$, $NE(jr,\ell,n)$, and $BACK$ (jb,ℓ,n) are computed given assumptions on availability data and start-up levels of the technologies. (See the *Note* in section C.6 above.)

The rationales for the constraints in C.7 are the same as those mentioned in the note at the end of C.6.

C.8. *Nuclear Stockpile Constraints for the Nuclear Fuel Cycle*

The nuclear fuel cycle is essentially the same as the production, delivery, and consumption sector discussed in C.1 through C.3. It adds one further feature, however, in that a full set of accounts is kept for nuclear materials.

$$\sum_{nt=1}^{n} \left[\sum_{jr} \left(\frac{fin(j,jr)}{nyear} + ann(j,jr) \right) ne(jr,\ell,nt) \right.$$

$$- \sum_{jr} \frac{fout(j,jr)}{nyear} ne(jr,\ell,nt-1) + \sum_{i\neq\ell} ntr(\ell,j,i,nt)$$

$$\left. - \sum_{i\neq\ell} ntr(i,j,\ell,nt) - \sum_{jj=1}^{5} M(i,j,jj,\ell,nt) \right] \leq 0$$

$$(j = 5, \ldots, 8; n = 1, 2, \ldots; \text{each } \ell),$$

where

$ann(j,jr)$ = annual inputs of resource j for operation of reactor type jr, metric tonnes per 10^{15} btu.

$fin\ (j,jr)$ = initial inventory of resource j in reactor type jr, metric tonnes per 10^{15} btu per period.

$fout(j,jr)$ = final inventory of resource j in reactor type jr, metric tonnes per 10^{15} btu per period.

$nyear$ = number of years per period.

The above inequality is the critical equation of the nuclear fuel cycle—the material balance equation. It states that the stockpile must be nonnegative for each of the using countries ($\ell = 1, 2$), for each nuclear material ($j = 5, 6, 7, 8$), and in each time period ($n = 1, 2, \ldots$). The last (fifth) term is the mining activity, relevant only for natural uranium and thorium. The third and fourth terms consist of exports and imports of nuclear materials from other regions, respectively.

The first two terms in C.8 represent the flows associated

with the nuclear fuel cycle itself. Each reactor is represented as having an initial inventory (*fin*), annual fuel recharge (*ann*), and a final fuel inventory (*fout*). For each operating reactor, track is kept of each of the three parts of the fuel cycle. Thus for a LWR operating in period 1, there must be an *initial* stock of enriched uranium. For each year there must be a fuel recharge. At the end of the period (e.g., the beginning of the second period), there is an outflow recorded according to the final inventory (details of the flows are given in table 3.5).

C.9. *Bounds on the Variables*

    C.9.a. Nonnegativity

All variables are nonnegative.

    C.9.b. Upper bounds

$$xc(\ell,m,mm,n) \leq UP(\ell,m,mm,n).$$

Upper bounds are imposed on the levels of consumption to approximate the demand function.

Constraint C.9.b states that actual consumption, xc, for each step of each demand function must be no greater than the upper limit of the demand interpolation, UP.

D. *Objective Function of the Linear Programming Problem*

    The objective is to minimize the costs minus the marginal utilities for each activity. Thus minimize:

$$\sum_{i,j,jj,k,\ell,n} c^x(i,j,jj,k,\ell,n)\, x(i,j,jj,k,\ell,n)$$

$$+ \sum_{i,j,jj,\ell,n} c^m(i,j,jj,\ell,n)\, m(i,j,jj,\ell,n)$$

$$+ \sum_{i,j,\ell,n} c^{ntr}(i,j,\ell,n)\, ntr(i,j,\ell,n)$$

$$+ \sum_{jr,\ell,n} c^{ne}(jr,n)\, ne(jr,\ell,n)$$

$$+ \sum_{\ell,n} c^{hyd}(n)\, hyd(\ell,n)$$

$$+ \sum_{jb,k,\ell,n} c^{bck}(jb,k,n)\, bck(jb,k,\ell,n)$$

$$+ \sum_{k,\ell,m,n} c^{xp}(k,m,n)\, xp(k,\ell,m,n)$$

$$- \sum_{\ell,m,mm,n,} utils(\ell,m,mm,n)\, xc(\ell,m,mm,n)$$

where

c^i = cost coefficients, $i = x, m, ntr, ne, hyd, bck, xp$, 10^9 dollars in 1975 prices per 10^{15} btu. Here and below, the superscripts $(x, m, \ldots)$ refer to activities listed in section B above.

$utils(\ell, m, mm, n)$ = utility associated with mm^{th} interpolation component of consumption in demand category m, area ℓ, and time period n, 10^9 dollars in 1975 prices of nonenergy goods equivalent per marginal "util." *Utils* are in terms of 1975 nonenergy consumption as numeraire.

The cost coefficients are:

$$c^x(i,j,jj,k,\ell,n) = \frac{1}{(1+r)^{n-1}}\{[(rx + \delta^{ex}(i,j))cap^{ex}(i,j,jj) + cur^{ex}(i,j,jj)] + royal(i,j,jj)(1+r)^{n-1} + tran(i,k,\ell) + tax(i,j,\ell,n) + \frac{1}{feff(j,k)}[(rx + \delta^{fuel})cap^{fuel}(j,k) + cur^{fuel}(j,k)]\},$$

$$c^m(i,j,jj,\ell,n) = \frac{1}{(1+r)^{n-1}}\{[(rx + \delta^{ex}(i,j))cap^{ex}(i,j,jj) + cur^{ex}(i,j,jj)] + (1+r)^{n-1}royal(i,j,jj) + tran(i,4,\ell)\},$$

$$c^{ntr}(i,j,\ell,n) = \frac{1}{(1+r)^{n-1}}\{tax(i,j,\ell,n) + tran(i,4,\ell)\},$$

$$c^{ne}(jr,n) = \frac{1}{(1+r)^{n-1}}\{(rx + \delta^{ne})cap^{ne}(jr) + cur^{ne}(jr)\},$$

$$c^{hyd}(n) = \frac{1}{(1+r)^{n-1}}\{(rx + \delta^{hyd})cap^{hyd} + cur^{hyd}\},$$

$$c^{bck}(jb,k,n) = \frac{1}{(1 + r)^{n-1}}\{(rx + \delta^{bck}(jb,k))cap^{bck}(jb,k)$$
$$+ cur^{bck}(jb,k)\},$$

$$c^{xp}(k,m,n) = \frac{1}{(1 + r)^{n-1}}\{[(rx + \delta^{xp})cap^{xp}(k,m)$$
$$+ cur^{xp}(k,m)] + markup(k,m)\},$$

where

$rx =$ cost of capital for investment decisions in energy sector, at a given rate per period.

$r =$ discount rate for future costs and benefits of the alternative technologies, at a given rate per period.

$\delta^i =$ depreciation rate, $i = ex, fuel, ne, hyd, bck, xp$, at a given rate per period.

$cap^i =$ investment requirement, $i = ex, fuel, ne, hyd, bck, xp$, 10^9 dollars at 1975 prices per 10^{15} btu capacity.

$cur^i =$ cost of current inputs, $i = ex, hyd, ne, hyd, bck, xp$, 10^9 dollars at 1975 prices per 10^{15} btu.

$royal =$ royalties on scarce energy resources.

$tran =$ transportation costs, 10^9 dollars at 1975 prices per 10^{15} btu delivered.

$tax =$ taxes on imports and exports, 10^9 dollars at 1975 prices per 10^{15} btu delivered.

$markup =$ markup of market price over marginal cost, including distribution and taxes, 10^9 dollars at 1975 prices per 10^{15} btu delivered.

For a derivation of the costs, see chapter 3. Also, note that the wage and capital costs, as well as the discount rate, reflect market prices or social opportunity costs depending on the exact case used.

The marginal utilities are derived from observed demand functions.

E. *Computer Implementation of the Linear Programming Model*

The preceding sections described, with optimal concision, the basic framework of the energy model used in the present work. To implement the equations, a computer program was created to write the linear program in the input format required by the programming code. The program aims at maintaining a high level of flexibility in order to accommodate different projections on future courses of energy production, conversion, transportation, and consumption. This flexibility leads to a capital intensive computer program—one that is extremely complicated and difficult to read but easy to use once (if?) mastered. An alternative specification of a few parameters will suffice to simulate a different scenario.

With this warning in mind, I refer the interested reader to the listing of the computer program generating the input for the MPSX, available on request in Nordhaus (1978). The usage of the code is described in the following IBM publications:

1. *Introduction to Mathematical Programming System—Extended (MPSX), Mixed Integer Programming (MIP) and Generalized Upper Bounding (GUB)* (GH20–0849–3).

2. *Linear and Separable Programming Program Description Manual* (SH20–0968).

F. *A Simple Example*

As the whole procedure may appear complicated and arcane, a highly simplified example of the exact procedure is useful. In the example we are confronted with the following situation:

1. There is only one demand category. Demand is completely price-inelastic and equal to $0.1Q$ ($Q = 10^{18}$ btu) per year. Either petroleum or solar energy satisfies demand with a 0.50 utilization efficiency.

2. There are two technologies:
    a. Petroleum has zero production and transport costs, but is limited to $2Q$.
    b. Solar energy is unlimited in supply, but total costs are $\$10^{12}$ per Q cumulative production.

3. The discount rate on goods is 10 percent per annum.

4. The program is calculated for 21 years.

To convert this problem to a linear programming format, define:

$x_1(t)$ = production of petroleum in year t,
$x_2(t)$ = production of solar energy in year t,
$d(t)$ = demand in year t.

The objective function is therefore

$$max\ z = \sum_{t=0}^{20} [10{,}000{\cdot}d_1(t) + 0{\cdot}d_2(t) - 0{\cdot}x_1(t) - 1{\cdot}x_2(t)](1.1)^{-t}$$

$$d_1(t) \leqq 0.1\ Q/\text{year}$$

In the first equation, the demand function has been divided into two steps, as illustrated in figure 4.1. The first step, $d_1(t)$, is that part of the demand function with very high marginal utility (10,000)—but $d_1(t)$ is constrained to be no greater than 0.1 Q per year, as shown in the last inequality, for $d_1(t)$. To represent the assumption that demand is very inelastic for the first step, a very large marginal utility is taken. Similarly, the second step, $d_2(t)$, has zero marginal utility, reflecting the fact that the demand is completely inelastic. The third and

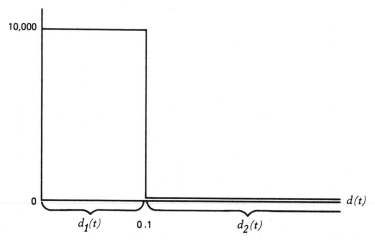

Figure 4.1.
Linearized form of objective function in simple problem.

fourth terms of the objective function are the cost of petroleum extraction and of solar energy. Finally, both marginal
utility and costs are discounted at 10 percent annually.

The remaining constraints are as follows:

$$x_i(t) \geqq 0$$
$$d_j(t) \geqq 0$$
$$0.5x_1(t) + 0.5x_2(t) - d_1(t) - d_2(t) \geqq 0, t = 0, \ldots, 20$$
$$\Sigma x_1(t) \leqq 2.0.$$

The first two equations are nonnegativity conditions. The
third equation states that production of petroleum and solar
energy, times their efficiencies (0.5), must be at least as great
as total demand. The last equation states that total extraction
of petroleum must be less than or equal to total resources (2.0).

The solution to the program is straightforward:

1. The $d_1(t)$ is always at its upper limit $[d_1(t) = 0.1]$
because it has such a high return, while $d_2(t) = 0$ because it
has no return and is costly.

2. Petroleum will be exhausted before solar energy is
produced, and the period of petroleum production runs
from year $t = 0$ through year $t = 9$.

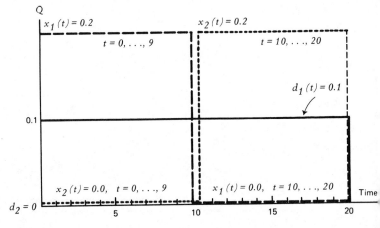

Figure 4.2.
Solution for activities in simple example.

A graphical demonstration of (1) and (2) is shown in figure 4.2. In addition as shown in Figure 4.3:

3. The shadow prices on the petroleum constraint (dual variable to the last inequality above) is $0.39 per million btu in period zero. This implies that the shadow price on each $x_1(t)$ is $0.39(1.1)^t$. Thus petroleum price starts at about two-fifths of the backstop price for solar energy and rises at 10 percent annually before equaling the backstop price at the date of exhaustion, year 10.

4. The shadow price for demand is just twice the shadow price of production because utilization efficiency is 50 percent.

A final point is that the program can be broken into a transition period (years 0 to $10 = \hat{T}$) and a stationary period (from year $10 = \hat{T}$ on). As long as the program goes at least to $\hat{T}$, the solution for the transition period is invariant to the horizon.

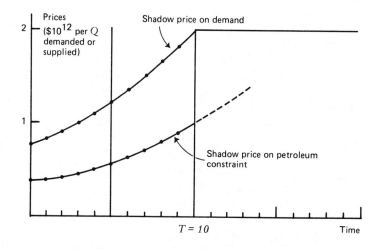

Figure 4.3.
Solution for dual variables ("shadow prices") in simple example (values are in constant rather than present value prices).

5   THE EFFICIENT ALLOCATION

OF RESOURCES OVER TIME

A. *Background*

A fundamental scientific and policy question faced by the United States and other countries is how the stock of exhaustible energy resources should be efficiently utilized. The stock of energy resources consists of some that are extremely inexpensive to recover but quite scarce (oil and gas, especially in the Mideast); others that are intermediate in cost and relatively abundant (coal, shale, and fissionable nuclear resources occurring in natural form); still others that are very expensive to convert to useful energy but are essentially inexhaustible (fertile nuclear material, fusion resources, solar energy, perhaps geothermal energy).

Although economic theory can tell us *qualitatively* how society should proceed to use its resources, it cannot determine a priori *how fast* the low cost resources should be exhausted. Or putting the question in terms of the relevant social decision variable, theory cannot tell us *how high to set the prices of different energy resources in order to assure their efficient use.*

The limited nature of the exercise of estimating efficient prices and quantities should be stressed at the beginning: the economic efficiency concept refers to paths which guarantee that the total real income of society attains its highest level. Put differently, an economically efficient path is one which has the property that it is not possible to raise the income of any nation or generation without reducing the income of another nation or generation. The economist's concept of efficiency is quite different from those of thermal or thermodynamic efficiency, which concern engineers.

In Nordhaus (1973) I presented preliminary results on the efficient allocation of energy resources over space and time. This chapter updates those estimates and adds several new analytical features that were not considered in the original. The new estimates differ in two basic respects: first, the data base upon which the estimates rest has been considerably revised in light of the post-1973 situation; and second, the treatment of information and capital taxes in the new version differs.

B. *Fundamental Assumptions*

In assessing the efficient use and pricing of energy resources, the following seven assumptions are basic to the estimating procedure:

B.1. *No Taxes on Energy Resources*

There are currently substantial taxes on energy resources, especially on oil exported in international trade since the rise of OPEC. In the estimates that follow, I treat all goods neutrally by setting the taxes on energy resources at zero. In doing so, however, I will estimate the efficient scarcity rents to all exhaustible energy resources.

B.2. *Environmental Policy*

One area where the evidence is deficient concerns the determination of an efficient environmental policy. For this reason, I take the environmental policy existing in 1975 as the relevant standards for the present purposes. The most important of these include: sulfur and particulate removal on all direct burning of coal (estimated to cost $40 per kilowatt thermal); emissions controls on automobiles; release and locational regulations on nuclear plants; restoration of lands used for surface mining (estimated to cost $5,000 per acre); safety regulations for deep mining.

The assumptions used here are deficient in two respects. First, because of the absence of a price mechanism in environmental goods, there is no assurance that prices of different components of environmental quality (mortality, morbidity, property damage, visual amenity, all in different regions) will have similar "shadow prices." For example, it has been argued that the shadow price of a statistical death prevented by routine releases in nuclear power is about 1,000 times more than the shadow price on a statistical death from coal-fired electricity. The presence of such anomalies indicate that the standards are inefficient. A second caveat with the present approach is that the "price of the environment" may well continue to rise in the future, where a rise would mean that the stringency of environmental standards would increase. Such an increase might lead to severe controls on burning of fossil fuels (as is considered in chapter 8 on carbon dioxide), on surface mining in general, on nuclear power or some aspect such as plutonium recycle, on the burning of coal, and so forth.

B.3. *Perfect Information and Long-run Equilibrium*

In the modeling that follows, the simulation assumes that the economy is in long-run equilibrium (although this will later be relaxed). Further, it is assumed that there are no uncertainties about future demands or technologies. It is recognized that the assumptions regarding perfect information and a full set of future markets do not realistically capture the nature of energy markets, in that there are no significant futures markets in the market economies.

B.4. *No Price Controls*

The efficient solution presented in this chapter assumes that there are no noneconomic controls on energy prices, or more precisely that prices are equal to marginal cost. This assumption is counter to a good deal of fundamental economic regulation, particularly in natural gas and electric utilities, in which the price is set at the average historical cost of production. It is well known that in competitive markets these controls or regulations have adverse effects on efficiency, even though they may have "desirable" effects on the distribution of income or wealth.

B.5. *Aggregation*

In the present chapter the level of regional aggregation is quite high, while the level of detail and temporal disaggregation is much greater. Unlike in the earlier study, Nordhaus (1973), it has been decided to aggregate the world into two regions, the United States and the rest of the world (ROW). The reason for the high level of regional aggregation is basically that for the estimation of efficient prices the transportation costs of energy resources are very low relative to the production costs, and therefore the errors in regional aggregation are relatively small. On the other hand, because of the importance of having fairly fine detail on certain technologies—especially the nuclear fuel cycle and synthetic hydrocarbons—much more attention has been given to those questions.

B.6. *The Discount Rate*

In general, the models used here assume that the market prices are the relevant "shadow prices" for different goods—this is true for labor, capital goods, materials other than energy, and so forth. In the case of comparing the value of goods over time, this procedure would use a *market interest rate* as a way of discounting future streams and of allocating exhaustible resources over time. This principle poses serious problems because pervasive market imperfections and taxes on capital and capital income lead to the existence of a multiplicity of different interest rates.

A particular difficulty is the existence of the corporation income tax, which effectively drives a wedge between the real social return (the pretax real rate of return) to capital and the real interest rate observed on markets (the post-corporation tax but prepersonal tax real rate of return). In addition, there are personal taxes on the nominal return from capital that further lower the full post-all-taxes real rate of return. Thus, after accounting for both the taxes on corporate capital and the personal income taxes, we obtain the discount rate that households would apply in making their decisions about whether or not to postpone consumption—the consumption discount rate.

In what follows I use two distinct approaches. In this chapter, in which I analyze the efficient use of energy resource, I assume that all distortionary taxes are absent and further that the post–all-tax rate of return to corporate capital is the relevant cost of capital. This leads to a discount rate on goods of 6 percent per annum, which is applied to both resources and to capital goods.

In the next chapters, in what I call the "market model," I am concerned with predicting the outcome of actual market processes and therefore include distortionary taxes. Therefore in chapter 7 I will use the pretax rate of return on capital, estimated to be 13 percent per annum, to calculate the cost of capital for investment and pricing decisions.

B.7. *Backstop Technology*

In the earlier study, Nordhaus (1973), considerable attention was given to the notion of a backstop technology as an important device for simplifying the estimation procedure. In the estimates that follow there are two different backstop technologies, electricity from nuclear fission in breeder reactors and central station solar electricity. Ideally, the estimates would be made by extending the terminal year until that point when the backstop technology has achieved 100 percent penetration of all sectors. Unfortunately, the limited computational budget allowed little experimentation with the horizon. It was discovered that the backstop technology had achieved 100 percent penetration after 12 ten-year periods, but for technical reasons stationarity has not been achieved.

A summary of the model is given in table 5.1. This shows in tabular form the fundamental assumptions used, as well as the time dimensions and the level of aggregation.

C. *Empirical Estimates of the Efficient Allocation of Energy*

I now present the empirical estimates of the efficient allocation of energy resources in the two-region world model. The model run reported here was for 12 ten-year periods, starting in 1970 and running through 2090. It was found that the backstop technology was reached in 2070, so (with a couple of caveats discussed below) there would be no change in the estimated path if the planning horizon were extended beyond

Table 5.1.
Basic Assumptions Used in the Efficiency Model Run

1. Time period:
   a. 120 years (1970 to 2090)
   b. 10-year steps

2. Regions:
   a. U.S.
   b. Rest of World (ROW)

3. Assumptions about OPEC:
   a. Aggregated in ROW
   b. Prices of oil are set efficiently (competitively)

4. Demand:
   a. Based on econometric estimates described in chapter 2
   b. Rest of World demand uses estimated demand functions but constant reduced 10% to account for aggregation error
   c. No short-run disequilibrium of demand functions (short-run elasticity equal to long-run elasticity)

5. Taxes:
   a. Current level of distribution and tax rates on energy services
   b. No taxes on energy resources

6. Resources:
   a. Assumes the base case of resource availability and cost functions as derived in chapter 3

7. Cost of capital and discount rate:
   a. All real costs of capital and discount rates are set at the estimated U.S. post tax yield on corporate investment, 6%

8. Environmental standards:
   a. Environmental costs include current U.S. legislation (through 1975) for emissions and air quality standards

9. Backstop technologies:
   a. Nuclear fission includes LWR, FBR, and HTGR technology as described in chapter 3
   b. Solar central station electricity generation used as described in chapter 3
   c. Assumed that inexhaustible nonelectric liquid or gaseous fuel available at $10 per million btu

10. Constraints on growth of new technology or production:
    a. None

2090. In what follows I will first discuss the processes that are chosen in the efficient program, then turn to the efficient prices, and finally discuss questions of sensitivity.

C.1. *Processes*

The first detail of the efficient solution is the set of least-cost technological processes. Table 5.2 shows the time path of processes for the United States over the planning horizon. Because of the way the processes are ordered, the *discounted* cost incurred in using backstop technologies is low at the left and bottom (as in using nuclear fuel for electricity generation in the distant future) and high at the top right (as in using electric or hydrogen cars right away).

The current model does not reach as complete a position of stationarity as in Nordhaus (1973) because of the revised treatment of the nuclear fuel cycle. True stationarity is reached when the breeders are producing so much plutonium or U–233 that they are free goods. Although stationarity did not occur in the 12 × 10 year run, plutonium does become a free good in the seventh and eighth periods of the 8 × 20 year run. Assuming there are no aggregation errors, this indicates that the position of true stationarity—all energy resources used being free or inexhaustible—occurs around 2090. In any case, the effect of this terminal period on today's decisions and prices is negligible, as is shown in section 5.E.

The pattern of interfuel substitution and the way in which processes unfold over time are sensitive to changes in parameters. The linearity of the objective function leads to extreme solutions. But while the exact details should not be taken terribly seriously, the efficient transition from exhaustible fossil fuels to the backstop technology shown in table 5.2 is the important result; and this basic pattern is insensitive to realistic uncertainties in costs.

Finally, figure 5.1 shows the total gross energy consumption for the United States. The actual figures for the period 1960 to 1975 are shown on the left-hand side of the graph, while the calculated are shown from 1975 to 2015. The calculated path starts out in 1975 at a slightly higher level than the actual path because the prices in the efficient path are lower than

Table 5.2.
Processes in Efficient Program, United States

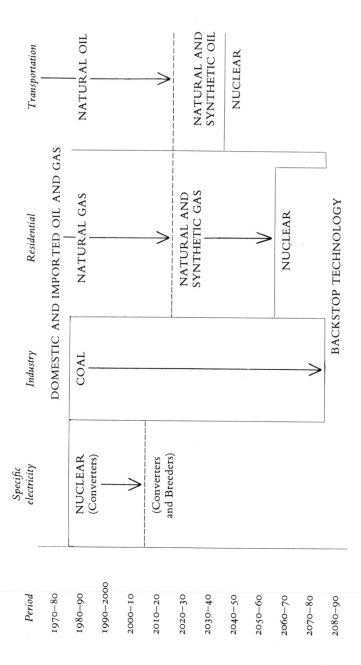

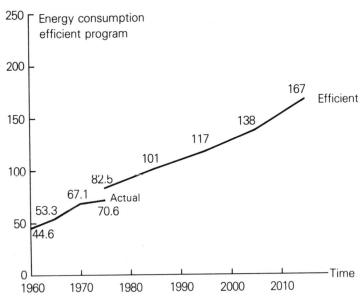

Figure 5.1.
Actual gross energy consumption, 1960–75, and calculated
1975–2015 efficient path (10^{15} btu per annum).

in the actual situation, as will be shown in tables 5.4 through
5.6. (The prediction along a path with actual prices will be
shown in later chapters.) However, even the efficient path of
energy utilization shows relatively modest growth in energy
consumption, with the path growing only 1.7 percent annually
over the period 1975 to 2000. The slow growth is principally
due to the fact that, as will be shown momentarily, the prices of
energy products are rising, in addition to the fact that the
energy demand is quite sensitive to the energy prices according
to the demand functions estimated in chapter 2.

C.2. *Prices*

Perhaps the most important economic question that is
addressed by the results is: What are the efficient prices on
scarce resources? The programming problem discussed in
chapter 4 estimates a set of "shadow prices" associated with the

solution. These shadow prices can be interpreted as the appropriate "rent" or "royalty" that a competitive market, operating with the same information we possess, would impute to scarce, low cost resources. Similarly, a world central planner, charged with attaining an efficient allocation, would find it essential to determine a set of "efficiency prices" to aid in allocating scarce resources.

Table 5.3 shows the shadow prices associated with the

Table 5.3.
Royalties (Shadow Prices) on Energy Resources, 1975 Prices

	1975	1985	2000	2050
Petroleum (per barrel)				
Drilled				
United States	$2.49	$4.46	e	e
ROW	1.71	3.06	e	e
Undrilled				
United States*	0.58	1.04	2.49	0.95
ROW	1.50	2.69	6.42	e
Coal, United States (per 10⁶ btu)				
Eastern Deep	0.00	0.01	0.02	0.34
Western Surface	0.02	0.04	0.08	1.56
Shale oil, United States (per barrel)				
25 gallon/ton	0.28	0.50	1.20	22.10
10 gallon/ton	0.02	0.04	0.09	1.63
Natural gas (per 10⁶ btu)				
United States*	0.36	0.12	0.29	1.02
ROW	0.14	0.25	0.54	e
Natural uranium (per pound)	1.31	2.35	5.62	103.6

NOTE: The royalties are the values of the dual variables on resources in the efficient solution. They exclude any future cost but include quasirents on past direct costs for drilled petroleum.

"e" indicates the resource is exhausted.

*Royalty pertains to the lowest-cost unexhausted resource and therefore need not rise at interest rate.

efficient solution. The most important result is that *the efficient royalty on oil and gas is far below the current world price. For the efficient source—low cost Rest of World oil—the efficient royalty is under $2 a barrel. This compares with the current royalty of about $10 a barrel.*

More generally, it is clear that in the efficient solution the scarcity shadow prices for resources-in-the-ground are modest. For gasoline, the royalty translates through to about $0.04 per gallon of gasoline (about 7 percent of current retail price in the United States). For coal, the shadow price is even smaller, being approximately $0.50 a ton for eastern U.S. coal. Natural gas has a shadow price of about $0.36 per million btu, while other fuels (shale and uranium) have miniscule scarcity rents.

It is also interesting to calculate the *fuel* prices that come out of the efficient solution. The fuel prices are the sum of shadow prices as well as the costs of extraction and correspond to the efficient market prices. Tables 5.4 and 5.5 show the time-path for prices for the United States in 1975 prices and compare these with actual prices over the past few years. The first point to note is that as of 1975, prices of all energy products except petroleum are relatively close to their long-run efficiency (or competitive) prices. This is particularly surprising given the fact that the estimates are based on engineering and geological data, drawn from many disparate sources, and given the enormous aggregation needed to obtain demand and resource categories.

Examining next the trend in calculated prices, we see divergent trends among the different fuels. The path for calculated baseload electricity prices shows almost no increase (0.1 percent annually over the next 35 years) as adaptation to a nuclear technology takes place. The calculated price of coal is also almost constant, rising 0.4 percent annually for 35 years. The time-path for petroleum and gas prices is much steeper, with calculated prices of crude oil and natural gas rising at around 3.7 and 2.3 percent annually, respectively. The reason for the projected run-up of petroleum and natural gas prices is that with the exhaustion of limited petroleum and gas resources we must turn to considerably costlier processes—either oil shale or

Table 5.4.
Intermediate Energy Prices Other than Petroleum, 1975 Prices

	Base-load electricity at busbar (mills per kwh)	Industrial Natural gas city gate (cents per 10^6 btu)	Industrial coal (delivered, dollars per ton)
Actual			
1950	24.9	12.8	9.39
1960	20.7	23.5	7.81
1970	14.3	23.8	8.72
1975	23.3	88.0	19.60
Calculated			
1975	10.5	88	20.20*
1985	11.1	112	20.40
2000	10.7	126	21.75
2010	10.7	194	23.50
2050	11.5	249	38.70
Annual percentage rate of change, 2010 over:			
1975 calculated	0.1	2.3	0.4
1975 actual	−0.7	2.3	0.5

SOURCES: Calculated values from the program as described in text. Electricity prices are from the Federal Power Commission, National Power Survey, 1970 for 1970; earlier years assume a constant real cost of transmission and distribution costs from 1970 and use the figures for the large industrial uses from Federal Power Commission, *Typical Electric Bills*. 1975 assumes the same real transmission and distribution costs. Natural gas coal and prices are from Bureau of Mines, Mineral Yearbook, various years. 1975 prices are from FEA (1976). Prices include direct costs and royalties. 1975 prices are obtained by using the GNP deflator.

*Supply price, but no production occurs.

Table 5.5.
Prices of Petroleum Products, 1975 Prices

Actual	Crude oil (delivered, dollars per barrel)		Refined oil products (wholesale, before taxes, dollars per barrel)	
	United States	World	United States	Western Europe
1950	5.08	n.a.	n.a.	n.a.
1960	4.99	3.54	n.a.	n.a.
1970	4.53	3.31	n.a.	n.a.
1975	10.38	13.93	11.34	12.45
Calculated				
1975	3.05	2.80	4.54	3.90
1985	4.40	4.20	6.04	5.80
2000	8.30	8.20	10.40	10.30
2010	11.10	11.20	13.40	13.60
2050	12.75	15.00	15.30	17.80
Annual percentage rate of change, 2010 over:				
1975 calculated	3.7	4.0	3.2	3.7
1975 actual	0.2	−0.6	0.5	0.3

SOURCES: Calculated values from program as described in text. Figures are from different sources for 1950–1970 and for 1975. For 1950 to 1970 actual, U.S. figures for crude oil are for beginning-of-year price of midcontinent crude (API, Petroleum Facts and Figures, 1971, p. 449). "World" crude price is from Adelman (1972), p. 365, where the figure for crude oil takes Adelman's realization less his calculated refinery margin. For product prices, these use U.S. product weights. Data for 1975 are from *Monthly Energy Review*, for the United States. Data for Western Europe from industrial trade journals for products, while crude oil for World uses the U.S. refiner acquisition cost of crude oil in 1975. Figures are converted to 1975 prices by applying the 1975 GNP deflator.

coal gasification and liquefaction. Thus in the efficient solution, and before any royalties, a barrel of oil can be delivered in the United States at $1.75 a barrel in 1975 while (at a 6 percent cost of capital) shale oil and liquefied deep coal cost about $10.00 a barrel. Before the technological transfer can be made from natural oil to synthetic oil, the price of petroleum products must rise very significantly.

The major discrepancy between the market prices and the actual prices are found for petroleum prices since 1973. Figure 5.2 shows the royalty (the calculated efficiency or competitive price) of ROW petroleum and compares this with the actual revenues flowing to producing countries. The clear evidence is that *up to around 1973 royalties were in the same general range as the calculated efficiency royalties.* In 1973, however, with the rise of OPEC, the royalties rose to approximately five times the efficient royalties. It is extremely tempting to explain the higher

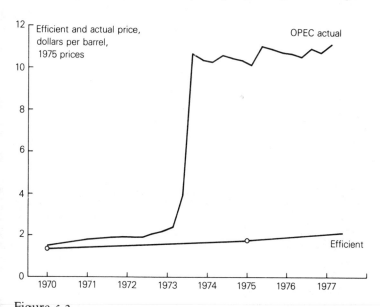

Figure 5.2.
Calculated and actual price of OPEC oil (Efficient and actual price, dollars per barrel, 1975 prices).

Table 5.6.
Prices per Million btu Delivered by Demand Category, 1975 Prices, United States

		Demand category		
Period	Delivered electricity	Delivered industrial	Residential*	Transportation[†]
Actual prices	7.96	1.66	3.15	24.05
Calculated prices[‡]				
1975	6.23	0.98	3.09	19.20
1985	6.55	1.31	3.49	20.40
1995	6.44	1.75	3.82	22.80
2005	6.44	2.11	4.39	26.20
2015	6.45	2.24	5.35	28.30
2045	6.52	2.90	6.61	28.20
2075 and on[‖]	6.85	6.40	8.03	28.70
Average annual percentage change 2015 over:				
1975 calculated	0.1	2.4	1.6	1.1
1975 actual	−0.5	0.8	1.3	0.4

SOURCES: Calculated values from program as described in text. These include direct costs and royalties. Actual from FEA (1976).

*Excludes electric.

[†] Gasoline only.

[‡] Figures correspond to tables 5.4 and 5.5 according to technology used in table 5.2. The differences in *levels* between tables 5.6 and the earlier tables are accounted for by the thermal efficiencies of different end uses and by taxes distributions. Thus the 1975 fuel prices for transport are 5.3 times the calculated gasoline price in table 5.5 to reflect the 19 percent efficiency of automobiles; in addition, they add $3.20 per million btu for taxes and distribution.

[‖] The price structure for the period 2075 on represents the prices associated with the backstop technology.

price since 1970 as the effect of monopolization of the oil market by OPEC, and the next chapter investigates that question in greater detail.

I next consider the prices of delivered energy. Table 5.6 shows the time-path of the prices for the four final demand categories. The story is roughly the same as in tables 5.4 and 5.5. The calculated rise for specific electricity is negligible, while for transportation the increase is modest. For industry and residential use the calculated rises are much larger—2.4 percent and 1.6 percent annually, respectively.

C.3. *Imports*

It will not be a surprise that the efficient path of energy allocation involves considerable imports into the United States during the early part of the period, particularly from 1970 to 2010. The estimates of the quantities and values of oil and gas

Table 5.7.
Oil and Gas Imports into United States, in Efficient Program,
1970–2010

| Period | Volume of imports of oil and gas (Million Barrels per Day Equivalent) | | Value of imports (Billions of 1975 dollars) |
	Million barrels per day equivalent	Percent of energy consumption	
Actual			
1975	6.9	21.	$28.1
Calculated			
1970–80	24.9	64.	27.7
1980–90	30.1	63.	48.4
1990–2000	36.6	66.	88.1
2000–10	20.9	32.	76.7
2010–20	9.4	12.	41.5

SOURCES: Calculated values are from program. Actual from U.S. Bureau of Census.

imports are shown in table 5.7, along with actual data for 1975. Although the balance of payments burden on the United States shown in the last two columns of table 5.7 is no larger than the current payment (actually it is slightly lower), the volume of imports is staggering. The annual imports in the efficiency calculation are 25 million barrels a day equivalent, as compared to the actual 1975 import volume of about 6.9 million barrels per day of oil equivalent.

D. *A Parable about Efficiency*

At the cost of repetition, it is worth digressing to consider a fallacy about the efficiency price of energy resources that has some casual intuitive appeal. It is sometimes claimed that the efficient price of oil should be equal to the price of its nearest substitute. The basic logic of efficient allocation is contrary to this view. In the present estimates, the closest substitute for oil is shale oil or liquefied coal, which costs approximately $10 a barrel (at a 6 percent cost of capital, higher at the market cost of capital as will be indicated in chapter 7). Yet the efficiency price of oil in 1975 is about $3 a barrel, only 30 percent of the substitute price.

What is the reason for this price relation? Basically it concerns the role of discounting and the productivity of capital. For example: if oil costs nothing to produce and it is priced at the cost of substitutes (a fair description for OPEC oil), this implies that many uneconomic investments will be made. The easiest way to see this is to pretend that capital is like robots, which can either mine and liquefy coal at a price of $10 a barrel or can reproduce and grow at a rate of 10 percent a year. A policy of pricing oil at the price of substitutes would mean that a society would be indifferent between buying cheap oil or diverting robots to liquefying coal. Say that one squadron of infinitely durable robots can make 1 million barrels per day of liquefied coal, and we have an initial stock of one squadron of robots. The inefficient policy is to have our robots make 1 million barrels of oil a day in perpetuity, which would be "economic" if oil is priced at the substitute price.

In an efficient economy, we would pump the (free) oil out of the ground as long as it lasts, say 30 years, and let the robots reproduce at the net rate of growth of robots, 10 percent per annum in our example. At the end of 30 years, we have 17.45 squadrons of robots, who can make 17.45 million barrels of synthetic oil a day in perpetuity. The efficient policy produces more oil in *every* period than the inefficient policy.

This modernized Aesop's fable has the important moral that, if capital has a positive marginal productivity, then this must be reflected in the efficiency price of scarce resources.

E. *Sensitivity of the Results*

The results of the model presented above represent a single model run, but they are the product of a very large number of assumptions. Although sensitivity runs are of great interest, they are also extremely costly. I have focused on the question of the sensitivity of the results concerning (1) the 1975 and 2015 efficiency price of natural or synthetic petroleum and (2) the long-run price of energy in the form of the 2055 price of electricity. The latter is chosen because electricity prices are the basic determinant of the price of energy goods in the all electric-hydrogen economy of the backstop technology assumed here. Finally, I have shown (3) the discounted value of real income originating in the energy sector in order to indicate the effect on real income of the different exogenous uncertainties.[1] It should be noted that to conserve computational resources, the base run for the sensitivity analyses has taken a 120-year period broken into 6 twenty-year periods rather than the 12 ten-year periods used above.

E.1. *Aggregation and Time Horizon*

The first question revolves around the temporal aggregation in the model. It was noted above that (certain minor problems aside) the calculation shows that the backstop technology is reached around 2070. This implies that the calculations should be invariant to the time horizons as long as it goes at least to 2070. In addition, we have tested the sensitivity of

---

1. "Real income originating" is the total value of consumption less the opportunity cost of substitutes, or, in technical terms, the value of consumer plus producer surplus.

Table 5.8.
Sensitivity Tests on the Intrinsic Scarcity of Energy Resources

	Price of crude oil (dollars per barrel, 1975 prices)		Production (busbar) cost of electricity, 2055 (mils/kwh, 1975 prices)	Change in discounted value of "real income" originating in energy sector, billions (1975 prices)
	1975	2015		
0. Base run for comparisons (six 20-year periods)	2.99	12.07	11.31	0
1. Aggregation and Time Horizon				
1.1. Twelve 10-year periods	3.05	11.76	11.46	−35,407
1.2. Eight 20-year periods	2.99	11.76	11.26	+11,769
2. Discount rate at 13% rather than 6%	2.20	13.73	19.35	n.a.*
3. Different undiscovered oil and gas resources				
3.1 50% lower	3.71	13.24	11.28	−1,446
3.2. 100% higher	2.90	11.76	11.08	515
4. Inelastic demand function	3.51	13.62	11.45	n.a.*
5. Differential technological change	2.90	6.43	2.55	35,843
6. Solar instead of nuclear "backstop technology"	3.13	13.83	28.08	−2,774
7. 1973 model and data	2.03	11.09	15.18	n.a.*

*These runs use different objective functions and therefore cannot be compared.

n.a. = not applicable

using a 10-year time step rather than a 20-year time step. Sensitivity run 1.1 in table 5.8 shows the effect of using the 10-year time step (the base run discussed above) on the solution. The overall effect is to change the shadow prices on both current and future energy goods by around 2 percent. However, moving to a longer time horizon (160 years in 20-year steps in run 1.2 as against 120 years in 20-year steps in run 1.1) makes no difference to the initial prices and only very slight differences to the final prices. The reason for the change in the final prices is simply that the program imputes zero value to the terminal stocks of nuclear materials, and therefore slight differences in the prices during the terminal periods will be observed. It is thus clear that the concept of a backstop technology is extremely helpful in choosing a terminal date, in that the prices are insensitive to changes beyond that date.

E.2. *Different Cost of Capital*

The efficient run makes use of a relatively low cost of capital (see the discussion in section 5.B.6). Many people would be more comfortable with a cost of capital more in line with the pretax cost of capital to corporations, around 13 percent. Therefore, I have made one sensitivity run with the higher cost of capital. The results of the higher discount rate are somewhat surprising: the current efficient scarcity price of petroleum *falls* from $3.00 to $2.20 per barrel, reflecting the greater "tilt" in the price path due to higher discounting of the cost at the switch to the substitute technology. On the other hand, the "futures" prices of both petroleum and the backstop technology are raised to reflect the higher rental cost of the capital equipment needed to produce the energy goods.

E.3. *Availability of Oil and Gas*

The figures used for the availability of oil and gas in this study are higher than many geologists accept, reflecting the assumption of considerable undiscovered resources; on the other hand, many economists feel that geologists are conservative in assuming there will be no major new fields or technologies discovered. From a review of the different work, discussed in chapter 3, it was found that the standard deviation of the resource estimates for oil and gas was approximately 13 percent of the mean. I have therefore run two separate estimates of the model with undiscovered oil and gas reserves 50 and 200 percent of the base case.

The outcome of the sensitivity run on the availability of undiscovered oil and gas resources is shown in lines 3.1 and 3.2 of table 5.8. The first column shows that if we are surprised by a 50 percent shortfall of the oil and gas resources, this would mean that the current efficiency price of oil should be about $0.70 per barrel (about 20 percent) higher in 1975, but that other prices are only slightly affected. If, on the other hand, the undiscovered resources were doubled, this has almost no effect on efficiency prices. The asymmetry in result is due to the fact that the royalty is equal to the discounted substitute price at the "switch" date, and that the switch date has a much bigger effect with a proportional move toward the present (in the case of the reduction) than toward the future (as in the case of the increase).

It is also interesting to note the effect on total discounted world income, shown in the last column. The uncertainty in oil and gas reserves amounts to about $2 trillion, or about one-third of the annual world GNP.

E.4. *Inelastic Demand for Energy*

One property of the model is that it has relatively price-elastic demand functions for energy products in comparison with other models.[2] I have therefore made a run in which the demand functions are very inelastic with respect to price, with final elasticities equal to −0.1 in all end use categories. This is shown in line 4 of table 5.8.

The effect of the inelastic demand functions is to make prices relatively higher, for in this case the demands are growing considerably more rapidly. The difference in the current efficiency price of oil is approximately 15 percent in the inelastic case.

2. See MRG (1977) and Taylor (1977).

E.5. *Differential Technological Change*

A central assumption of the model is that the rate of technological change in the energy industry will be at the same rate as the rest of the economy. This is contrary to long-standing historical trends, in which energy costs have fallen around 2 percent relative to total GNP [see Kendrick (1961)]. Therefore, I have included a run in which there was differential technological change at the rate of 2 percent annually.

Of all the changes, the differential productivity growth makes the most difference for the future and the least for present decisions. The 1975 scarcity price of petroleum falls about 5 percent, while the scarcity price for future (2015) petroleum falls almost 50 percent and the backstop price of electricity in 2055 falls to 25 percent of its base value. Moreover, the present value of the continual technological change is enormous, $36 trillion of discounted real income.

E.6. *Alternative Backstop Technologies: Solar Instead of Nuclear Power*

A major uncertainty in future energy systems is the question of whether a backstop exists and, if so, what it will be. It is fairly well agreed among energy specialists that, as far as resource availability is concerned, there are two backstop technologies, which are proven in small-scale but not large-scale deployment: nuclear fission breeders and solar technology. The ultimate costs and the environmental acceptability, especially of the nuclear technology, are still open questions. Because of the uncertainties in the ultimate technology, it therefore seems worthwhile to investigate the possibility that the nuclear technology will not be acceptable and that it will be replaced by another backstop technology.

I therefore made a further run *omitting the possibility of using nuclear technology*, but allowing the use of solar technology (as in fact did all earlier runs). The price of solar [derived from the CONAES study which supported MRG (1977)] was assumed to be approximately twice that of the nuclear breeder technology, more precisely $1730 per kilowatt electric converted to a 75 percent load factor, or $800 per peak kilowatt at the estimated 35 percent load factor.

The results of the backstop sensitivity analysis were somewhat surprising: there was only a 5 percent increase in the scarcity price of petroleum (less on all other 1975 prices, not reported here) and a 15 percent increase in the 2015 petroleum price. Electricity prices eventually rise 150 percent over the base case, a very substantial margin. The last column shows the gross economic cost, the discounted value of world real income originating in the energy sector, before any calculations about the costs or benefit of externalities (increased use of coal in the early part of the period, reduced risk from nuclear power throughout the period, and increased risks from solar toward the end of the period). The gross cost over the period of not allowing any nuclear power is estimated to be $2.8 trillion, or about 2 percent of total discounted world GNP over the entire period.

E.7. *Comparison with Earlier Estimates*

Perhaps the best way to test the sensitivity of the estimates is with respect to the earlier runs presented in Nordhaus (1973), shown in line 7 of table 5.8. The current estimate of the efficiency price of petroleum has been raised approximately 50 percent over the 1973 estimate (both converted to 1975 prices). The price of electricity, on the other hand, is approximately 25 percent lower than the earliest estimate.

There has been no very detailed attempt to track down the difference in the estimates, but there are two major factors that account for the change. First, the lower cost of capital (interest rate) in the current run is the major element responsible for the lowering of the backstop price, as can be seen by comparison of lines 0 and 2 in table 5.8. For the petroleum price, on the other hand, the differences are, first and most important, that a lower discount rate is used (as can be seen by comparing lines 2 and 0); and, second that there has been a major shift in the relative price of discovering new oil and gas reserves and in producing synthetic fuels such as shale oil or liquefied and gasified coal. Since the cost of the alternative resource to natural oil and gas looks much higher than earlier estimates, the scarcity value of petroleum rises pari passu.

6 A Quantitative Estimate of Market Power in the International Oil and Gas Market

A major departure from reality of the efficiency model of chapter 5 is in the assumption that free trade between different economic regions prevails. Especially since the rise of OPEC and the dramatic price rises of 1973, it has been widely believed that pricing of petroleum products does not reflect competitive conditions. The results of the last chapter (see especially figure 5.2) reinforce this belief and indicate that the current market price of oil is approximately five times the efficiency or competitively determined price.

Many explanations exist for why the price of oil is not equal to its competitive level, but the major hypothesis is surely that the price is largely determined by OPEC, and that OPEC is largely interested in obtaining high, if not maximal, profits from its reserves. More technically, it is hypothesized that OPEC behaves so as to maximize the present value of the profits from sales on its oil. Such behavior on the part of a monopolist has been analyzed theoretically in chapter 1. This chapter gives numerical results that flesh out that theoretical framework.

The structure of this chapter is the following: section A provides the details on the estimation procedure used here, while section B asks, what would the limit price for OPEC be if only the United States were involved? Section C presents estimates for the world, and section D compares these with earlier estimates.

A. *Computational Aspects*

The problem of estimating the wealth-maximizing price for OPEC requires deeper techniques than the mathematical programming algorithms used up to now. In principle, a fixed-point algorithm (such as the Scarf algorithm) could estimate these prices, but at present the size of the wealth-maximizing oil-price problem is far too large for economical use of fixed point algorithms. Therefore, we must rely upon the theoretical insights derived in chapter 1 to help make an approximate calculation of the wealth-maximizing prices.

A.1. *Analytical Approach*

Analytically it was shown in chapter 1 that the monopolist's problem can be formulated as follows: The monopolist can act as a price setter, setting the price of oil in each time period. Subject to these prices, then, consumers react passively by consuming oil in a pattern that maximizes their satisfaction.

Unfortunately, the real-world problem is much more complicated than the simple one-good model investigated in chapter 1. In the model used here, there are several time periods, two destinations, four monopolist's resources (two grades each of oil and gas), and four demand categories. The techniques available at this time do not allow the exact determination of the wealth-maximizing price trajectory in such a large problem.

A simpler approach can be used, however, to derive an upper bound or *limit price* for monopolized oil and gas. When the model is solved with no production of the monopolist, the prices which result are the outer limit on what will be obtained by the wealth-maximizing monopolist. As shown in chapter 1, in certain simple cases the wealth-maximizing price will be very close to the limit price.

A.2. *Modification of the Model for Monopoly Runs*

In what follows, I will generally prefer model runs that have made one special change for the monopoly problem. The basic difficulty with the standard competitive model used above is that the short-run demand functions are extremely price inelastic, with elasticities for the first period (up to 1980) of approximately −0.05. Because of the extreme inelasticity, the monopoly and limit prices for the first, and sometimes the second, period rise so high as essentially to erase national income. This is especially the case for the two-region model (U.S. and ROW), as will be shown in section C of this chapter.

The difficulty arises from using partial equilibrium techniques in which the demand elasticities are bounded away from −1. This difficulty has been noted in several places, and was brought to my attention by Tjalling Koopmans, along with a suggestion for resolving the difficulty.[1] The procedure used here is a minor modification of the Koopmans procedure.

To implement the Koopmans variable elasticity suggestions, I have made three assumptions, as illustrated in figure 6.1. First, the elasticities that were observed in the econometric results presented in chapter 2 are assumed to hold in the neighborhood of the observed prices for 1970. This is the solid line in figure 6.1, while the dots schematically represent observations. Second, a unit-elastic demand function (e.g., one with the observed income and population elasticities, but with its own price-elasticity of −1) is drawn through the 1970 observation of price and quantity. This is shown by the dashed line in figure 6.1; note that it actually goes through the 1970 observation.[2]

Finally, the demand function actually used in the calculations, labeled the "new demand function" in figure 6.1, is one obtained by taking a linear combination of the "observed demand function" and the "unit-elastic demand function." Thus at P_{1970}, the new demand function is identical to the

1. Koopmans' suggestion was contained in Koopmans (1974) and later discussions.

2. This is the major difference between the Koopmans proposal and the present realization of that proposal: Koopmans proposed that unit-elastic curves lie above and below the observed demand curve by fractions k and $1/k$; and that this pair of curves would be the actual demand curve whenever the inelastic curve lay above or below, respectively, the unit elastic curves.

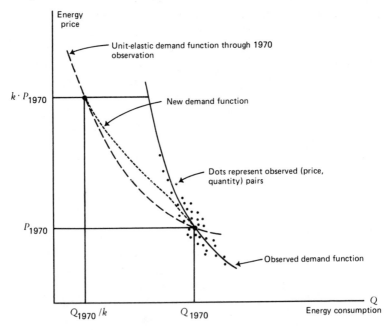

Figure 6.1.
Illustration of the technique used to modify the demand function so as to prevent more than total income being spent on energy products in the case of extreme price rises.

---

observed demand function; halfway to $k \cdot p_{1970}$, the new demand function is midway between the two functions, while at and above the price $k \cdot p_{1970}$, the new demand function coincides with the unit-elastic demand function.

This technique will affect the outcome only for the first 10 or 20 years of the estimation period. During these years, the price elasticity is very small because of assumed lags in response. After 1990 there is no essential difference, for after that point the demand functions become quite elastic.

B. *Computational Results for the United States*

First asked is a relatively simple question: What are the limit prices and wealth-maximizing prices when the only im-

Table 6.1.
Basic Assumptions Used in the OPEC Model Run

1. Time period:
    *a. 60 years (1970 to 2030)
    b. 10-year steps

2. Regions:
    a. U.S.
    *b. Rest of World
    *c. OPEC (oil and gas exports only)

*3. Assumptions about OPEC:
    a. Determine limit price

4. Demand:
    a. Based on econometric estimates described in chapter 2
    b. Rest of World demand uses estimated demand functions but constant reduced 10% to account for aggregation error

5. Taxes:
    a. Current level of distribution and tax rates on energy services
    b. No taxes on energy resources

6. Resources:
    a. Assumes the base case of resource availability and cost functions as derived in chapter 3

7. Cost of capital and discount rate:
    a. All real costs of capital and discount rates are set at the estimated U.S. post tax yield on corporate investment, 6%

8. Environmental standards:
    a. Environmental costs include current U.S. legislation (through 1975) for emissions and air quality standards

9. Backstop technologies:
    a. Nuclear fission includes LWR, FBR, and HTGR technology as described in chapter 3
    b. Solar central station electricity generation used as described in chapter 3
    c. Assumed that inexhaustible nonelectric liquid or gaseous fuel available at $10 per million btu

10. Constraints on growth of new technology or production:
    a. None

*Indicates modifications from chapter 5.

porting region considered is the United States? This is a problem that is both simpler computationally and for which the data of the model are more carefully adjusted than for the world.

B.1. *The Standard Run*

The *standard run* used is described in table 6.1. It is one in which I have limited the length of the time period considerably, to only 6 ten-year periods. The reason for this limitation is that the monopoly problem mainly concerns the next 30 years, the period during which the economy is constrained on both the supply and demand sides by inadequate or inappropriate capital equipment in several parts of the energy sector. By the year 2000, sufficient adjustment of both production and consumption should have occurred, so that the monopoly price of OPEC oil will be essentially the cost of the next cheapest alternative energy source. But until 2000, there will be an additional premium on OPEC sources because new sources or conservation practices have not yet been put into place.

One significant caveat must be noted: It is assumed that there is marginal cost pricing of all energy resources. This assumption implies that there are no historical cost pricing practices, or price controls, in any sectors—an assumption that, of course, violates reality for all parts of the energy sector.

The first line in table 6.2 gives the results of the standard run. These are extremely interesting for their implications about the role of imports in the U.S. economy. The first point is that, in a full short-run limit-price equilibrium—that is, one where energy goods are priced at marginal cost rather than average cost—the general level of prices is considerably higher than in the efficient case. The average price is approximately $2.32 per million btu as opposed to the *actual* average value for 1975 of $2.04 per million btu and an average price in the efficient case of $0.50 per million btu. The *composition* of the price differential is striking. The actual and no-import price for oil are in fact very close, this in part reflecting the fact that average oil prices are today quite close to marginal costs. Prices for natural gas, by contrast, are well below the btu-equilibrating price; after equilibration, the price for gas in the no-import situation is 300 percent higher than the actual price.

Table 6.2.
Results of Sensitivity Analysis for Limit Prices for Oil Imports

		Limit price for oil (dollars per barrel, 1975 prices)			
		1970–80	1980–90	1990–2000	2000–10
1. Standard run:					
4 × 10 years		9.62	14.45	14.64	13.63
2. Pessimistic oil and gas runs: growth in drilling limited to 4% and delay 20 years on synfuels		8.87	17.86	18.60	20.76
3. Short-run price elasticity sensitivity runs:					
	short-run				
k-value	elasticity*				
a. $k = \infty$	−0.07	35.94	13.23	14.64	13.62
b. $k = 20$	−0.30	11.25	14.44	14.64	13.62
c. $k = 0$	−1.20	7.83	14.33	14.23	13.62
4. Subsidy on oil and gas consumption (per million btu): $0.50 oil; $1.00 gas		10.69	14.20	14.64	13.63
5. Different temporal aggregation					
a. 6 × 10 years		9.62	14.75	15.20	15.60
b. 8 × 5 years		9.81	15.78	15.96	14.55
6. Efficiency price (see chapter 5)		3.05	4.40	8.33	11.06

NOTE: All sensitivity runs used 4 ten-year periods.

*"Short-run" is for an average lag of 5 years, calculated from 1970 price to two times 1970 price for *delivered* energy for transportation sector.

The significantly higher average price level is responsible for the lower levels of consumption. Even with the very high price levels, due to the low price elasticity, consumption declines only 16 percent from the current level.

The major finding in the standard case, however, is that the *limit price on OPEC oil is fairly modest*. Table 6.2 shows the limit price over the next 40 years in the standard program. The limit rises from a current value of $9.62 per barrel, as high as $15.60 a barrel in the fourth decade—all in 1975 prices. The prices in the next two decades (not shown) decline.

There are three serious reservations to this finding—two reflecting economic uncertainties, one reflecting shortcomings in the modeling. (1) The results reflect what some experts and models find to be "optimistic" assumptions about the rate of expansion of the domestic oil and gas industry. (2) The demand elasticities are highly uncertain and are higher than many believe plausible. (3) The model abstracts away from the fact that market prices are well below the marginal costs in table 6.2. The effects of each of these uncertainties are investigated in the next section.

B.2. *Pessimistic Oil and Gas Assumptions*

A major uncertainty about future energy policy concerns the response of U.S. domestic oil and gas production to the higher prices since 1973. The uncertainties concern the ultimate resources that will be available, the marginal cost of these resources, the rate at which drilling will grow in the near future, and, finally, the extent of market power in the oil industry.

Of these, the most important for the question of the limit price on imports are the costs of extraction and the maximum rate of growth of drilling. In the standard program, the extraction cost function rises in several steps, with a marginal cost of $6.70 per barrel for the first 20 billion barrels, $7.80 per barrel for the next 20 billion barrels, followed by $9.43, $12.00, $16.00, and thence upward for each 20 billion barrel increment. Approximately the same costs hold for natural gas on a btu basis. Similarly, it is assumed that until 1978, the rate of growth of drilling will be − 10 percent annually for natural gas and − 2 percent annually for oil, while after that period the rate of growth of drilling will be no greater than 7 percent annually for both.

To test the sensitivity of the results to the technological assumptions about the oil and gas industry, I have made the following further run: first, I have decreased the maximum rate of growth of drilling in the post-1978 period from 7 percent to 4 percent; second, I delayed the introduction of synthetic fuels by 20 years.

The results of the oil and gas sensitivity run are shown in

table 6.2. At the top of the table is shown the results of the standard run, discussed in section 6.B.1. On line 2 is shown the results of the pessimistic oil and gas sensitivity run. This indicates that the main effect is not on the first period (when production is pretty much predetermined anyway) but on later periods. By the fourth period, 2000–2010, the limit price is slightly more than 50 percent higher than the standard run.

Thus, if our assumptions are too optimistic with respect to the U.S. oil and gas industry, this may give too "flat" a slope to the limit price; in the "pessimistic oil and gas" case, the annual rate of growth in the real limit price is approximately 3 percent annually, as against 1 percent annually in the standard case.

B.3. *Different Assumptions about Price Elasticities*

The major determinant of the limit price in the short run is the extent of price inelasticity in the demand for energy. In the standard run used above, I have used the econometrically estimated elasticities, with a k-factor of 8. This implies that when energy prices rise by a factor of 8, the arc-price elasticity from the 1970 level is unity.

There is clearly considerable uncertainty about the demand elasticities, although it is clear that the short-run price elasticity is very small [see Taylor (1977)]. I have performed a certain amount of experimentation, however, to see whether different assumptions about the k-factor make a substantial difference to the results. The third section in table 6.2 shows the results of these experiments. It is important to note that the level of the k-factor only influences significantly the first-period limit price on oil, and that from the second period on there is sufficient elasticity in both demand and supply that the limit price has settled down into a fairly narrow margin. Thus the range for the first period price ranges from $7.83 per barrel when $k = 0$ through $9.62 per barrel for the standard run when $k = 8$ (shown in line 1 of table 6.2), to $35.94 per barrel when $k = \infty$. For the second period, on the other hand, the values range from $13.23 per barrel to $14.44 per barrel. A similarly narrow range extends for periods three and four.

The results of the demand elasticity sensitivity analysis

indicate that there is considerable uncertainty about the short-run limit price, even in the case where prices equal marginal cost. Once demand and supply responses have come significantly into play, however, the short-run inelasticity question is no longer significant.

B.4. The Effect of Prices That Are Not Equal to Marginal Costs

The final sensitivity question addressed is the role of markets for which prices are market-clearing—i.e., the fact noted in section 6.B.1 that prices are well below estimated marginal costs. This will be called the problem of "energy subsidy." It is extremely difficult to attempt to estimate the effects of an energy subsidy in mathematical programming models, but I will give a rough attempt. Assuming that an industry is regulated so that it charges average rather than marginal cost, this will under certain circumstances be equivalent in a mathematical programming model to a subsidy on energy consumption of an amount equal to the difference between average and marginal costs. I therefore estimate the effect on the market outcome when energy consumption in different fuels is subsidized.

A rough estimate of the degree of subsidy is obtained by comparing the marginal and average cost of fuels. In 1975, according to FEA, the difference between marginal cost and average cost per million btu is $0.73 for natural gas and $0.61 for oil.[3] In the run made here, I use a subsidy of $1.00 per million btu for gas and $0.50 per million btu for oil. The results of the subsidy-sensitivity run is shown on line 4 of table 6.2. As can be seen, the effect of the subsidy is relatively minor, with an increase in the import limit price of $1.07 per barrel in the first period and an actual decrease in the second period. No differences are seen in the later two periods.

In summary, I have found that the limit price for oil imports into the United States has considerable sensitivity to the assumptions about the structure of the economic model used. In the short-run, up to 1980, the major question is the

3. FEA (1976).

short-run price elasticity of demand. The range of estimates for the limit price is between approximately $8 per barrel in the case where there is no difference between the short-run and the estimated long-run elasticity to $36 per barrel when the short-run elasticity is consistent with a linear phasing in of new capital equipment. These compare with the efficiency estimates in the present work, derived in Chapter 5, of $2.70 per barrel.

For the longer term, there are a number of uncertainties, of which the most important is the question of how fast the U.S. oil and gas industry can expand and how quickly and at what price synthetic fuels come on line. The range of prices for the period 2000–10 is from slightly above $13 per barrel to about $21 per barrel (all in 1975 prices). Although no extensive sensitivity analysis is performed on this result, it is clear that the cost of oil and gas and of synthetic fuels is a most important variable in determining the limit price for the later periods.

C. *Estimates for the World Model*

The present model is not carefully calibrated to estimate the limit and wealth-maximizing price for the two-region (U.S. and ROW) model, but it is interesting to give very crude estimates in this case. In making estimates for the world model, I have run the following experiment: first, I assume that in 1973 the non-U.S. oil production is divided into OPEC and non-OPEC sources. Capacity for the non-OPEC sources outside the United States is estimated to be 22 million barrels a day in 1976.

Next, to calculate the limit price, it is assumed that production of OPEC is simply set at zero, while the U.S. and non-OPEC production behaves competitively. The production constraints on oil and gas are, first, that the rate of growth of production cannot exceed 7 percent annually; second, that the production must come from the non-OPEC capacity. These are estimated to be 28 percent of oil and 57 percent of gas for 1975. The production structure for the rest of the world is simpler than that for the United States in that there are only two grades of petroleum and of gas and there is no separation between the drilling and production decisions.

The results are shown in table 6.3. These indicate that the level of wealth-maximizing prices for the entire world is extremely uncertain, especially for the early part of the period. I have shown a standard run, in which the k parameter is equal to 8 as in the standard run for the United States, as well as the same set of sensitivity runs as was performed for the United States.

For the standard run, the limit prices for oil start out quite low at $11.80 per barrel for the current decade, but then rise rapidly as demand adjusts and as non-OPEC sources come on line. The figure for 1980–90 is $21.27 per barrel, while the decades following are $23.85 and $24.70 per barrel (always in 1975 prices).

I have not engaged in extensive sensitivity analysis except for the question of the short-run elasticity. Table 6.3 shows the effect of different short-run price elasticities on the time-path of the OPEC limit price: the short-run price elasticity referred to is the elasticity of demand for fuel in the transportation sector—this being almost exclusively petroleum products.

Table 6.3.
*Limit Prices in World Model: Equilibrium Prices When No OPEC Production Occurs**

Runs for different short-run price elasticities: k-value	Short-run[†] arc price elasticity	Limit price for oil, dollars per barrel, 1975 prices decade centered on:			
		1975	1985	1995	2005
$k = 4$	−1.1	9.45	18.60	26.45	25.00
[‡]$k = 8$	−0.6	11.80	21.27	23.85	24.70
$k = 16$	−0.4	15.40	27.15	23.65	24.15
$k = 50$	−0.2	25.60	30.85	21.00	24.15
$k = \infty$	−0.07	82.80	34.50	18.19	24.15

*Prices in ROW approximately 2% higher than U.S.

[†]"Short-run" is for an average lag of 5 years, calculated from 1970 price to two times 1970 price for *delivered* energy for transportation sector.

[‡]Standard run for U.S. takes this k-value.

The market runs reported in other chapters in the present work use a k-factor of infinity, indicating that the first-period elasticity is *very* small, equal to -0.07 in the transportation sector. For this very inelastic case, the OPEC limit price (i.e., the price that would lead to no net exports by OPEC in the period 1970–80) is an average price of more than $80 per barrel; this constituting a 25-fold increase over the 1970 level. If the demand function has a k-factor of 50—which corresponded to a price elasticity of -0.2 over the range of 1 to 2 times the 1970 levels for the demand for fuel in the transportation sector—the limit price is considerably smaller, about $25 per barrel in 1975 prices. For k-factors in the range from $k = 0$ to $k = 16$, the limit price is in range of $7 to $16 per barrel.

This result should be interpreted with some caution, however, for it seems extremely unlikely that the price for the current decade will be anywhere near $15. If the nominal price rises at 10 percent for the rest of the period through 1979, the average price over the decade will still average only approximately $8 per barrel, which is still well below the limit price for any but the most optimistic elasticity estimates.

For future periods, the model used here gives much less uncertain estimates about the limit prices. The range of figures for the 1980–90 period is $18 to $34 per barrel for the limit price, while for the 1990–2000 period the price range is $18 to $26 per barrel. Finally, for the first decade of the next century the estimate for all k-values is in the order of $25 per barrel. It should be emphasized that the very narrow range in later periods merely shows the effect of different k-factors and omits the larger uncertainties for that period, including such questions as the uncertainty about the growth of demand, synthetic fuel prices, and non-OPEC oil and gas resources.

D. *Comparison with Other Estimates*

To gain some perspective on the analysis presented above, the results are compared with other models.[4] In general, the present work differs from previous models in three respects.

---

4. This section was prepared with the assistance of Richard Peck.

First, the work discussed here is substantially more disaggregated than its predecessors—taking into explicit account, for example, detailed processes, resources, and interfuel substitution among oil, gas, coal, and nuclear power. Second, the treatment uses empirically estimated demand functions. Finally, previous models use an optimization approach, whereas this model estimates a set of limit prices.

Table 6.4.
Techniques and Size of Models Used to Estimate Optimal or Limit Prices for OPEC

Author	Number of regions, demand	Number of regions, supply*	Time period	Type of model	Computational algorithm
Pindyck (1977)	1	2	1975–2010	Dynamic optimization	Nonlinear optimal control
Hnyilicza and Pindyck (1976)	1	3	1975–2010	Dynamic optimization	Nonlinear optimal control
Ezzati (1976)	10	2	1975–1985	Simulation	Linear programming
Cremer-Weitzman (1976)	1	2	1975–2015	Dynamic optimization	Modified gradient method
Kennedy (1976) and Houthakker (1976)	6	4	1980	Static	Quadratic programming algorithm
Kalymon (1975)	2	2	1975–2000	Dynamic optimization	Classical constrained optimization
Blitzer, et al. (1975)	1	2	1974–1995	Simulation	Computer simulation
Nordhaus	2	2	1970–2010	Dynamic limit pricing	Linear programming

* All models except Nordhaus consider explicitly only one fuel, petroleum.

OPEC pricing models fall into three categories, a consequence of attempting to answer separate though related questions. The dynamic optimization models determine a price path over time that would maximize the sum of discounted future profits of the OPEC cartel. The static models address the stability of the OPEC cartel in maintaining prices in the medium run. Falling midway between the optimization and static models, the dynamic simulation models serve to assess the viability of OPEC rule-of-thumb price scenarios. Table 6.4 gives a general overview of the differing models and techniques used to estimate OPEC optimal or limit prices.[5]

The results are summarized in figure 6.2 and table 6.5. The bottom line of the figure shows the competitive or efficiency price, which was approximately in effect before the 1973 price rise. Most of the studies of OPEC price behavior indicate that the wealth-maximizing price would lie between $10 and $20 per barrel (1975 prices) over the next decades. The only exception to this is the limit price with very inelastic demands, which starts at a very much higher price but then declines to below $20 per barrel in the mid-1990s.

These studies confirm the view that the price rise of the early 1970s can be traced to the effective monopolization of the international oil markets. They also indicate that, given the very inelastic short-run demand for energy, the world oil price can show very great movements above the long-run monopoly price over a period of a decade or more without hurting producer revenues. It would appear, however, that any rise in the oil price significantly above the range of $10 to $20 per barrel (1975 prices) would take prices well above the long-run monopoly price.

5. A more complete exposition is available on request.

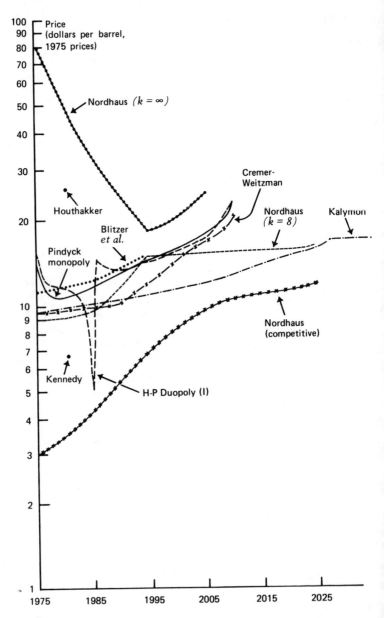

Figure 6.2.
Estimates of OPEC optimal or limit price, alternative models,
1975 prices, (for key, see text and tables 6.4 and 6.5).

Table 6.5.
Optimal, Limit, or Equilibrium Prices for OPEC under Alternative Models, 1975 Prices

	1 P	2 H-P-I	3 H-P-II	4 K	5 C-W	6 B	7 K	8 H	9 N-E	10 N-S	11 N-∞
1975	14.47	16.23	15.73	9.48	—	11.37	—	—	3.04	9.00	82.81
1980	10.80	11.53	11.31	9.83	9.80	11.81	6.69	25.50	—	—	—
1985	11.85	5.08	12.33	10.17	—	12.35	—	—	4.40	9.62	34.50
1990	13.10	13.33	13.53	10.49	10.30	13.55	—	—	—	—	—
1995	14.41	14.32	14.78	10.93	—	—	—	—	7.02	14.75	18.19
2000	15.81	16.09	16.09	11.26	14.70	14.98	—	—	—	—	—
2005	17.40	17.62	17.62	11.75	—	—	—	—	9.70	15.20	24.15
2010	22.18	22.23	22.23	12.57	20.80	—	—	—	—	—	—

KEY: 1 Pindyck monopoly (P)
2 Hnyilicza-Pindyck duopoly with varying production ratios (H-P-I)
3 Hnyilicza-Pindyck duopoly with fixed production ratios (H-P-II)
4 Kalymon (K)
5 Cremer-Weitzman (C-W)
6 Blitzer et al. upper limit price (B)
7 Kennedy: revenue maximizing OPEC tax (K)
8 Houthakker: revenue maximizing OPEC tax (H)
9 Nordhaus efficient run (N-E)
10 Nordhaus OPEC price limit standard run, $k = 8$ (N-S)
11 Nordhaus OPEC limit price, $k = \infty$ (N-∞)

7 ENERGY ALLOCATION
WITH MARKET IMPERFECTIONS

A third application of the energy model is now presented—
one that estimates energy consumption and prices in a realistic
market environment. Unlike either the polar extremes of
the perfectly efficient markets of chapter 5 or the wealth-
maximizing monopolist of chapter 6, this chapter attempts to
capture the imperfections of real world economic systems. It
is therefore necessary to address many of the actual problems
of the world energy market: market power, differential taxation, inefficient provision of new technology, lags in the
response of production and consumption to price signals, and
so on.

In section A, the assumptions of the market modeling are
outlined with a comparison to econometric techniques, and
then the major inefficiencies that are built into the model are
discussed in detail. Section B presents the results for the two-
region model (U.S. and ROW). These results can then be
compared with the efficiency and monopolistic models of
chapters 5 and 6. Finally, section C gives an analysis of the
economic costs of the inefficiencies in the market system.

A. *Assumptions in the Market Model*
    The market model uses a novel technique to simulate the
time-path. It is an "optimization" model—that is, the calculation of the path uses mathematical programming (generally

linear programming) to calculate prices and quantities. It should be emphasized that the technique of optimization does not in any way imply that the outcome *is* optimal. Rather, the mathematical programming algorithms are ways of finding the equilibrium solution in a market where many of the economic agents are optimizing.

As described in chapter 4, the technique followed is to maximize "utility" subject to technical and other constraints. In mathematical programming models, there is an automatic computation of shadow prices associated with each constraint, along with the quantity path for each variable.[1] The *shadow price* for each resource, intermediate flow, or end use of energy is the derivative of the attained maximum with respect to the availability of that good from outside the system. In optimization models, the two paths—one of "optimal" quantities and one of shadow prices—are the counterpart of the quantity-price projections of econometric models. The shadow price concept, however, also applies to situations where market prices are hard to visualize. In these cases, the shadow price concept can be used to make up for the fact that the objective functions usually take account only of benefits from goods and services bought by the consumer, and neglect environmental variables. The shadow price associated with a constraint that curtails or eliminates the use of a technology measures the economic cost of that constraint. For instance, a shadow price associated with an environmental standard for maximum allowable concentration of CO_2 in the atmosphere can be used to estimate the overall economic cost of that standard, as in chapter 8.

In every respect the market model uses the same technological data as the efficiency model: there are no differences in the geological data or the cost data for different technologies even though the market structures are different. Rather, the pricing of goods and factors differs, as does the speed with which new technologies are developed and deployed. The specific differences in assumptions between the efficiency and

---

1. This point is stressed in MRG (1977), which also contains a fuller discussion.

market models are: the availability of nuclear and other new technologies; the presence of export taxes by OPEC; the presence of constraints on the rate of growth of production in both new and old technologies; the lack of anticipation of future price developments by consumers; and the presence of taxes on income from capital. The differences will be discussed one by one.

1. In the efficiency runs, it was assumed that there were no delays on the development of new technologies; if R&D costs were relatively low, such an assumption would be appropriate for the efficient allocation of resources. In a realistic market environment, however, it is clear that new technologies

Table 7.1.

Dates and Levels of Introduction of New Technologies

Type of technology	Date	Level
Electric		
Nuclear		
LWR	1970	6 GW^e
LWR—Pu recycle	1979	6 GW^e
LMFBR	2001	6 GW^e
Other types*	Never	
Nonnuclear		
Solar or fusion†	1989	6 GW^e
Nonelectric		
Coal gasification	1989	0.4 quads/yr
Coal liquefaction	1995	0.4 quads/yr
Shale oil	1995	0.4 quads/yr
Electric automobile	2025	Any level
Thermal hydrogen	2030	Any level

SOURCES: See chapter 3.

*HTGR, HWR, and advanced LMFBR introduced in section C of this chapter, but not in base market run.

†In what follows, both solar and fusion are taken to be alternative backstop technologies. The best of these is assumed to win, and the entry represents the best of solar or fusion.

are expensive and often tend to be developed too little and too late.

For the market runs, I have taken account of this phenomenon by assuming that technologies can be introduced only with a substantial delay, and even then they are subject to a growth constraint of approximately 10 percent per annum (slightly higher in early years). Table 7.1 shows the rates of introduction, as well as the levels of introduction, of important new technologies.

2. From both a technological and an economic point of view, there are limitations on the growth of existing technologies. From studies of penetration of technologies—see Fisher and Pry (1970)—it is seen to be rare that technologies grow at rates faster than 15 percent annually. This is especially true of activities such as energy production, in which there are serious bottlenecks and physical limitations in the short run [see especially FEA (1976)]. The assumptions used are shown in table 7.2.

3. It is clear that one of the major sources of inefficiency is the presence of monopoly power in the oil and gas market. Chapter 1 above has given a theoretical treatment of this question, and Chapter 6 has provided very rough estimates of the range of wealth-maximizing prices. As it was not possible to find a robust set of estimates for the wealth-maximizing or limit prices, it was decided to use current level of OPEC prices. Thus, I assume that the price of OPEC oil and gas, exclusive of scarcity rent, would be $13.30 per barrel (or $2.30 per million btu) in 1975 prices over the indefinite future.

4. In the efficient runs, we have assumed that consumers

Table 7.2.
Limits to the Annual Growth Rates of Production

Activity	Maximum Growth Rate
Shale oil, uranium and thorium mining, all nuclear technologies, liquefaction and gasification of coal, solar, fusion	10% annually
Coal mining, oil and gas drilling	7% annually

correctly anticipate the future trends in prices of energy products. As a result, the consumer demand functions would be the long-run demand functions. In the market model, such an assumption is unwarranted: the energy crisis of the 1970s clearly caught consumers and producers by surprise.

In the market runs, therefore, we assume that the world was on the long-run demand functions in 1970, and in 1975 consumers were way off their long-run demand functions. The exact representation of the demand functions is that the demand for energy depends upon specific capital equipment. The capital equipment is assumed to have a 20-year lifetime in all sectors. Thus if the price rises occurred in 1973, and a three-year lag of consumers to the price rise is assumed, the full reaction to the rise in prices would occur by 1996. In the interim period, the elasticity of demand is assumed to be a linear function of the number of years that have elapsed between 1976 and 1996.

5. The final inefficiency we investigate is the fact that income from capital is taxed, often heavily taxed, in most developed economies. The market model uses market prices in its simulations. In a world where one price for goods exists, this procedure is relatively straightforward: for capital, it would imply using *the* market interest (or discount) rate. The presence of capital taxes makes the simple procedure invalid. The corporation income tax, for example, effectively drives a wedge between the real social return (the pretax real rate of return) to capital and the real interest rate observed on markets (the post corporation tax but pre-personal tax real rate of return). In addition, personal taxes on the nominal return from capital further lower the full post-all-taxes real rate of return. After accounting for both the taxes on corporate capital and the personal income taxes, we obtain the discount rate that households would apply in making their decisions about whether or not to postpone consumption—e.g. the consumption discount rate.

In this chapter, we are interested in predicting the outcome of market forces, with all the imperfections, distortionary taxes, and monopoly power that occurs in these markets.

Therefore, I will estimate the results with the taxes in place, and the market prices and quantities reflecting these distortions. It is estimated that the pre-tax real rate of return on corporate investment is 13 percent, while the post-all-taxes real discount rate on consumption is 6 percent. (See MRG (1977) for a fuller treatment.) In addition, for simplicity of implementation, we have used the 6 percent discount rate for exhaustible resources.

B. Empirical Estimates of Market Allocation of Energy Resources: World Model

Now presented are the empirical estimates of the market allocation of energy resources in the two-region world model. The basic model run reported here is for 12 ten-year periods, starting in 1970 and running through 2090. In all respects other than the changes noted in section A, then, the computational details are the same as those for the efficiency runs of chapter 5. A summary of the assumptions is given in table 7.3. In what follows, I will first discuss the processes that are chosen in the market program, then turn to the simulated market prices.

B.1. Processes

The first detail of the market solution is the set of chosen technological processes. Table 7.4 shows the map of processes in the market runs for the United States, this map being comparable with table 5.2. The dominant technologies used in each time period and for each of the four demand sectors are shown in this table. It is interesting to note that the general pattern of utilization of technologies is quite similar to that for the efficient utilization; I will only discuss the substantial differences. In the specific electric sector, there is no use of oil and gas for generation in the market runs, whereas the efficient runs had oil and gas used in the first decade. Instead, the market runs use both coal and nuclear. For the next four periods, from 1980 to 2010, both the efficient and the market runs envisage use of nuclear converters, such as the light water reactor, to generate all base-load electricity. Transition to breeders occurs sooner and more quickly in the market runs,

Table 7.3.
Basic Assumptions Used in the Market Model Run

1. Time period
   a. 120 years (1970 to 2090)
   b. 10 year steps

2. Regions:
   a. U.S.
   b. Rest of World (excluded in section C of this chapter)

3. Assumptions about OPEC:
   a. Aggregated in ROW
   *b. Real prices of OPEC oil and gas (exclusive of scarcity rents) are set at levels of 1975, $2.30 per million btu

4. Demand:
   a. Based on econometric estimates described in chapter 2
   b. Rest of World demand uses estimated demand functions but constant reduced 10% to account for aggregation error
   *c. Short-run disequilibrium of demand functions (short-run elasticity not equal to long-run elasticity) based on 20-year lives of capital equipment

5. Taxes:
   a. Current level of distribution and tax rates on energy services
   *b. No taxes on energy resources other than OPEC export taxes

6. Resources:
   a. Assumes the base case of resource availability and cost functions as derived in chapter 3

7. Cost of capital and discount rate:
   *a. Posttax real costs of capital and discount rates on investment, and pretax return on resources, are set at the estimated U.S. posttax yield on corporate investment, 6%
   *b. Pretax rate of return on investment set at 13%

8. Environmental standards:
   a. Environmental costs include current U.S. legislation for emissions and air quality standards

9. Backstop technologies:
   *a. Nuclear fission includes only LWR and FBR technology as described in chapter 3
   b. Solar central station or fusion electricity generation used as described in chapter 3
   c. Assumed that inexhaustible nonelectric liquid or gaseous fuel available at $10 per million btu

10. Constraints on growth of new technology or production:
    *a. Constraints on dates of availability of new technologies
    *b. Rates of growth of all production limited to between 7 and 10 percent annually

*Indicates modifications of assumptions from chapter 5.

Table 7.4.
Processes in Market Program

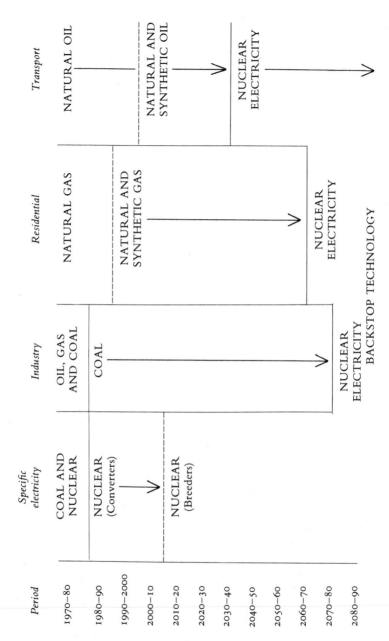

Period	Specific electricity	Industry	Residential	Transport
1970–80	COAL AND NUCLEAR	OIL, GAS AND COAL	NATURAL GAS	NATURAL OIL
1980–90	NUCLEAR (Converters)	COAL		
1990–2000				
2000–10	NUCLEAR (Breeders)		NATURAL AND SYNTHETIC GAS	NATURAL AND SYNTHETIC OIL
2010–20				
2020–30				
2030–40				
2040–50				
2050–60		NUCLEAR ELECTRICITY BACKSTOP TECHNOLOGY	NUCLEAR ELECTRICITY	NUCLEAR ELECTRICITY
2060–70				
2070–80				
2080–90				

because advanced converters are assumed not to be available, while in the efficient runs the mixture of advanced converters and breeders persists for some time into the future.

In the industrial sector, the patterns of utilization are identical: in both cases, coal is the most desirable fuel for process heat; for this reason, it is used exclusively after the first decade. There is also no substantial change in the fuel used in residential use; the major difference is that the use of natural oil and gas is much longer in the efficiency case, reflecting the much lower costs of imports. In the market case, synthetic oil and gas assume a substantial share of the market in 1990–2000, while for the efficient run the use of synthetic gas is not necessary until 2020.

Finally, in transportation the same general pattern is seen, with synthetic oil assuming a significant market share approximately 30 years earlier in the market model than in the efficient case. In both cases, the backstop technology is reached in the twelfth period, 2080–90.

In summary, the imposition of realistic market constraints on the world energy system is estimated to have substantial changes in the pattern of energy allocation over time or space. As would be expected, the major effect would be to accelerate the introduction of substitute fuels for those that are monopolized.

B.2. *Total Energy Consumption*

The next question addressed is the effect of the market constraints on the levels of energy consumption. In general, we would expect that the imposition of these constraints would lower the demand for *energy services* (i.e., for services such as delivered heat and passenger miles). It is not always clear, however, that the total gross energy inputs, commonly called *energy consumption*, would also decline, since the decrease of services might simultaneously take the form of shifting to fuels or processes with lower thermal efficiency.

Table 7.5 reports the total energy consumption for the United States over the twelve periods for both the efficient and the market runs, with the average annual growth rates

Table 7.5.
Total Energy Consumption, United States, in Efficiency and Market Runs

	Energy consumption in quadrillion btus	
	Efficient	*Market*
Actual		
1950	34.0	
1960	44.8	
1970	67.1	
1975	71.0	
Projected		
1975	82.5	71.9
1985	101.4	86.6
1995	117.6	97.0
2005	138.3	132.0
2015	167.2	170.2
2025	212.6	209.8
2035	227.6	233.2
2045	283.8	246.2
2055	319.4	278.7
2065	344.2	291.4
2075	399.2	243.1
2085	517.8	384.4
Average annual percentage rate of growth:		
Actual		
1950–75	3.0	
Projected		
1975–95	1.8	1.5
1975–2015	1.8	2.2
1975–2045	1.8	1.8
1975–2085	1.7	1.5

shown at the bottom. The first and most important point is
that the *level* of consumption in the initial period is significantly
higher, approximately 15 percent higher, in the efficient case
than in the market case. This reflects the lower prices in the
efficient case, as well as the more rapid adjustment of consumption and production to market prices. The second point
is that by the end of the twentieth century, the difference in
the level of the paths has disappeared. Indeed, *the total energy
consumption is actually higher for the market path than for the
efficient path in the period centered on 2015.* This paradoxical
result comes from the fact that the market path induces a
great deal of substitution of fuels that are much less thermally
efficient, especially synthetic liquids and gas in the middle
periods. By the end of the period under investigation, however,
the growth of consumption in the market path has slowed
below that for the efficient path, so that by the end, total
consumption is only 74 percent of the efficient path. The
reason for the slower growth in the later part of the period
is that most of energy consumption is based on nuclear electricity, and this has a higher price due to the higher cost of
capital in the market runs, thereby inducing lower demand.

Finally, one important implication of the present work is
that growth in energy consumption over the foreseeable future
is projected to be significantly slower than over the past few
decades. Historically, energy growth in the United States has
been approximately 3.5 percent annually (this being the figure
for 1950 to 1970). In the future, it seems unlikely that the
demand growth will be in excess of 2 percent unless the fundamental assumptions of the present work are incorrect. The
reasons for the slow growth are two: in the medium term,
up to 2010 or so, the moderate growth in demand is due to
the effects of significantly higher energy prices, which induce
lower demand. After 2010, when energy prices have generally
stabilized, the moderation is due to a slowdown projected for
the growth of aggregate GNP. Taken together, these two
forces give projections of demand consumption growth in
the range of 1.5–2.2 percent annually for subperiods in the
next 120 years.

B.3. *Prices*

One final important question addressed by the market runs is the extent of difference in the prices of the market and efficient runs. The first question is the level of shadow prices on exhaustible resources. Table 7.6 shows the shadow prices for important resources and compares these with those estimated in the efficient runs.

There are no major surprises in the results. All the competitively supplied (i.e., non-OPEC) resources have an increase

Table 7.6.
Shadow Prices for Energy Resources, Efficiency and Market Runs, 1975

Resource	Shadow price	
	Efficiency run	*Market run*
Petroleum (per barrel)		
Drilled		
United States	$2.49	$9.28
OPEC	1.71	0.18
Undrilled*		
United States	0.58	2.45
OPEC	1.50	0.10
Coal United States (per 10⁶ btu)		
Eastern deep	0.000	0.006
Western surface	0.020	0.020
Shale Oil, United States (per barrel)		
25 gallon/ton	0.28	0.31
10 gallon/ton	0.02	0.02
Natural gas (per 10⁶ btu)*		
United States	0.36	0.52
OPEC	0.14	0.03
Natural uranium (per pound)	1.31	2.30

*Pertains to the lowest-cost inexhausted resource.

in shadow prices—these corresponding to the increases that would occur in the market prices of these resources-in-the-ground. The most dramatic increases are those resources that are most competitive with the OPEC oil and gas—thus the shadow price on U.S. drilled petroleum increases about $6.75 per barrel, while the tax on OPEC oil is approximately $12 and the market price of oil increases $9.00 per barrel. Other resources that are less close substitutes increase more modestly, with very little increase in the shale and coal royalty (reflecting the superabundance of these resources). The rise on natural gas is quite modest, reflecting only a small fraction of the rise in oil prices; this is due to the fact that OPEC oil is projected to make little penetration into the U.S. gas market due to high transportation costs.

The shadow prices on OPEC oil and gas are quite interesting, in that they *fall* after the imposition of the export tax on oil and gas is imposed. It is quite astounding, then, that OPEC oil has an *internal shadow price* that is in the order of $1.75 per barrel, as compared to the $12 per barrel export price. The internal shadow price would be the efficiency price that OPEC countries should charge *themselves* in making their domestic investment and allocation decisions. This rate would be much lower than the export price to reflect the fact that their own opportunity cost of oil does not include the tax. In the case of natural gas, the internal shadow price is even lower.

There has been little discussion of the implications of a dual pricing system by OPEC countries, with the high monopoly price being charged for export, but the much lower internal shadow price being used for domestic investment. Recalculation of rates of return on investments that are intensive in the use of oil and gas, especially electricity generation, fertilizer, and hydrocarbon projects, suggests that the OPEC countries might well specialize in these sectors with a great deal of the lower costs passed on to consumers in the competition between different countries. This would, of course, represent a real lowering of the monopoly cost of oil and gas, but in ways that are hidden from cartel control.

Next examined are the product prices for the market and efficiency runs. These are shown in table 7.7, which is in the

Table 7.7.
Calculated and Actual Intermediate Energy Prices, United States, 1975 Prices

	Electricity at busbar (mils per kwh)		Industrial natural gas, city gate (cents per 10⁶ btu)		Crude oil (delivered, dollars per barrel)		Refined oil products (wholesale, before taxes, dollars per barrel)	
Actual								
1950	24.9		12.8		5.08		n.a.	
1960	20.7		23.5		4.99		n.a.	
1970	14.3		23.8		4.53		n.a.	
1975	23.3		88.0		10.38		11.34	
Calculated		Efficient Market		Efficient Market		Efficient Market		Efficient Market
1975	10.5	28.3	88	278	3.05	12.93	4.54	16.00
1985	11.1	17.9	112	273	4.40	12.78	6.04	15.90
2000	10.7	17.6	126	312	8.30	13.80	10.40	17.00
2010	10.7	17.9	194	291	11.10	14.30	13.40	17.60
2050	11.5	18.3	249	372	12.75	18.00	15.30	21.60
Annual percentage rate of change, 2010 over:								
1975 calculated	0.1	−1.3	2.3	0.1	3.7	0.3	3.2	0.3
1975 actual	−2.2	−0.8	2.3	3.5	0.2	0.9	−0.5	1.3

SOURCES: Calculated values from the program as described in text. Actual from tables 5.4 and 5.5.

same format as the corresponding table for the efficiency runs in table 5.4. Table 7.7 shows the results for products other than coal. These show product prices substantially higher in the market than in the efficiency runs: for 1975, for example, the price of electricity is 169 percent higher in the market case, while natural gas is 215 percent and coal (not shown) is 108 percent higher. Petroleum shows even greater differences, with the market run having a price 325 percent higher than the efficiency run.

There are several reasons for the market runs showing higher prices than the efficiency runs. A substantial part of the increase for 1975 is due to the export taxes of OPEC, which raised the market price of OPEC oil by the full 325 percent shown in table 7.7. This has induced effects on the other fuels, although, as for royalties, the effects are considerably smaller for nonpetroleum products.

In the longer run, even after OPEC oil and gas is effectively exhausted, the prices of fuels are higher than in the efficiency runs. Part of this, especially around the year 2000, is due to the slow buildup of synthetic fuels. Most of the difference, however, is the fact that the cost of capital for the market runs includes the effects of capital taxation and is 13 percent, whereas in the efficiency runs the cost of capital is only 6 percent (the latter, it will be recalled, corresponding to the assumed posttax supply price of capital). The long-run difference is most clearly seen in the case of electricity, where the price rise is 59 percent higher for the market case.

C. *Economic Costs of Market Inefficiencies in the Allocation of Energy Resources*

Section B examined the characteristics of a "market solution" to the allocation of energy resources. This solution differs from the efficient allocation of energy resources because of the presence of forces such as monopoly power, slow adjustment of consumers, and lags in the introduction of new technologies. The purpose of the present section is to examine briefly the *economic costs of inefficient allocation*. The property of an efficient solution is that it allocates energy resources (1) to provide the appropriate amount of energy services and (2) to produce the given amount of energy services efficiently. As an example of inefficiency of the second kind, consider a situation where the demand for energy is completely inelastic, so that the quantity of energy services is constant. In the presence of monopoly power, the price of oil might be raised so much that an alternative source of energy were used—say highly capital intensive liquefied coal were used as a substitute. As a result, the total amount of resources spent to attain the same level of energy services would be higher in the inefficient case.

In the programming models used here, it is relatively easy to measure the net economic losses stemming from most of the inefficiencies that are discussed above. The technique for estimation is to examine the change in the attained maximum value of the objective function between the efficient and the market solutions. As long as the change between the runs is due to changes in the technology (such as a given process coming in later) or as long as the accounting is carefully treated (so that the OPEC revenues are added back into the attained values), then the change in the value of the objective function is the appropriate measure of the change in the discounted value of real income. The only inefficiency that *cannot* easily be estimated in this way is the effect of inelastic demand functions.

In the estimates that follow, I first estimate the effect of all of the inefficiencies in the market run, then estimate them one by one.

C.1. *Overall Estimate*

Table 7.8 presents the overall results for the introduction of five inefficiencies simultaneously:

1. Limitation on the reactor configuration to the Light Water Reactor and the Liquid Metal Fast Breeder Reactor.

2. Constraints on the rate of growth of production due to the lack of forward markets to plan future production efficiently.

3. Constraints on the rate of introduction of new technology due to the inefficient provision of new technical knowledge.

Table 7.8.
*Losses in Real Income between Market and Efficiency Runs: Value
of Real Income, Discounted to 1975, 1975 Prices, Billions of
Dollars**

	Non-OPEC regions	OPEC taxes	World
Change in real income from efficiency run of:			
Efficiency run	0	0	0
Market run	−14,793	10,478	−4,315
Loss in discounted real income in market run:			
Total	−14,793	10,478	−4,315
As percent of discounted GNP	4.3%	—	1.2%

*See note to table 7.9.

---

4. Constraints on imports into the United States, limiting
the ratio of imports to GNP to the 1975 level.

5. Presence of taxes on OPEC oil exports constant in real
terms at the 1975 level.

For further discussion of these inefficiencies, see section A of
this chapter, and the discussion therein.

The overall loss in efficiency is staggering. According to
the estimates, the inefficiencies in the energy market account
for a reduction of approximately 4.3 percent in the discounted
future income of the non-OPEC countries and approximately
1.2 percent of the discounted income of the world. The non-
OPEC regions lose a discounted value of $14.8 trillion (dis-
counted at 6 percent), while the OPEC countries gain approximately $10.5 trillion in tax revenues; the difference, $4.3 trillion,
is the deadweight loss to the world economy as a whole. Put
differently, even if the OPEC revenues were recycled back to
those consumers who were hurt by the inefficiencies, only
71 percent of the loss could be made up by the recycling.

C.2. *Estimate of the Effect of Individual Components*

The estimates of the economic costs of the different sources of inefficiency in section 7.C.1 can also be broken into separate components. In making the estimates, I have made the following runs: First, I make the "base market run," in which all five inefficiencies listed in section 7.C.1 are included as constraints. Then I estimate the economic value of removing each of the five constraints (actually four different runs). By comparing the value of real income originating in the energy sector in the two runs, I can estimate the cost of each inefficiency using the base market run as a starting point.

Table 7.9 shows the results of this analysis. Along the left-hand side of the table are the run names, starting with the efficiency run and the base market run, then showing four variants of the base market run. Next shown is the level of discounted GNP (approximately calculated) for each run, where this is divided into non-OPEC, OPEC, and the world total. Finally, the last two columns show the losses in the particular run from the efficiency run (i.e., the inclusion of the three or four inefficiencies) and the gain in real income compared to the base market run.

The results indicate that the major individual inefficiency is the presence of export taxes on oil and gas. With no export taxes and no constraints on imports, the discounted value of world income is $3.3 trillion higher. Of this, approximately $14 trillion represents a gain by non-oil-producing countries, while $10.5 is loss by oil-producing countries, for a net gain of $3.3 trillion. Putting this loss on an annualized basis, it represents approximately 1.0 percent of total income or of the discounted value of GNP.

The second most important inefficiency is the presence of growth constraints on production; this amounts to $1.8 trillion of discounted income. An examination of the details of the run indicates that almost all the costs are due to constraints on the growth of non-U.S. oil and gas production, which is far below the optimum that would arise as a result of the OPEC export taxation.

The third most important inefficiency is the inefficient

Table 7.9.

Calculations of Discounted Real Income and Efficiency Losses for Individual Components of Market Run, Real Income* (Billions of Dollars, 1975 prices)

| | | | | Loss or gains | |
Run	Non-oil producing regions	Oil producing regions	World	Loss over efficiency run	Gain from market run
1. Efficiency	343,000	4,000	347,000	0	4,315
2. Base Market Run (BMR)	328,207	14,478	342,685	−4,315	0
3. BMR with all reactors allowed	328,238	14,478	342,716	−4,284	30
4. BMR with no production growth constraints	329,619	11,269	340,888	−6,112	−1,797
5. BMR with no constraints on new technologies	328,652	13,719	342,372	−4,628	−313
6. BMR with no import constraints or export taxes	342,035	4,000	346,035	−965	3,350
Total loss					4,315
Sum of individual components					5,491

*Note that the attained value of the objective function has no intrinsic meaning as it includes a constant of integration. I have therefore added a constant term such that the attained value of the objective function is equal to the discounted value of GNP.

Finally, I have taken GNP to be approximately $6.2 trillion for 1975. Discounted at a rate of 6%, the discounted value of world GNP is $347 trillion. OPEC discounted GNP is taken to be $4 trillion in the efficient case.

timing of new technologies. The source of the loss is that the growth of nuclear technologies is slower than would be efficient in the light of the rise in fossil fuel prices after 1973. The loss due to slow introduction of new technologies is about $300 billion in discounted value.

The final inefficiency, and one which is rather small relatively, is the limitation on the use of alternative reactor configurations. The use of the advanced converter reactors, intermediate and high-gain HTGRs, is especially heavy in the rest of the world because of the constraints on the development of the breeder and constraints on plutonium in the base market run. Notwithstanding, this inefficiency costs only $30 billion of discounted real income.

One bizarre and unanticipated result appears in the calculations shown in table 7.9. Note that the base market run has the worst level of real income for the non–oil producing regions, but the same is not true for world income. The result is an application of the general theory of "the second best," which states that when all the marginal conditions for efficiency do not hold, moving toward satisfaction of a single marginal condition may lead to a deterioration in the overall level of welfare.[2] *Relaxation of the growth constraints on new technologies (line 4) leads to a deterioration of world income, rather than an improvement, by a total of $1.8 trillion.* This strange result is a reflection of the fact that the world social cost of substituting new technologies is greater than the private cost. In this case the loss of income to the oil producing countries that results from the expansion in the competitive production more than offsets the gains in income to non–oil producing countries, and this net negative increment leads to a very substantial decline in world income. A similar, but less dramatic, decline with relaxation in a constraint is seen for the case where there are no constraints on the growth of new technologies.

2. See Lipsey and Lancaster (1956).

8 STRATEGIES FOR THE CONTROL

OF CARBON DIOXIDE

A. *Climatic Effects of Energy Use*

In recent years, the concerns about the trade-off between economic growth and environmental quality have become central to economic policy. One less-well-known example is that emissions of carbon dioxide, particulate matter, and industrial heat may, at some time in the future, lead to significant climatic modifications.[1] This chapter investigates the possibility of major climatic effects of energy use over the next century due to combustion of fossil fuels. In addition, alternative strategies for controlling carbon dioxide are discussed.

A.1. *Energy and Climate*

Climate usually refers to the average of characteristics of the atmosphere at different points of the earth, such as temperature, precipitation, snow cover, and winds. A more precise representation of the atmosphere is as a dynamic, stochastic system of equations. The probability distributions of the atmospheric characteristics is what is meant by *climate*, while a particular realization of that stochastic process is what is called the *weather*.

Recent evidence indicates that, even after several millennia,

---

1. See especially Matthews et al. (1971), Kellogg and Schneider (1974), and Schneider (1976).

STRATEGIES FOR THE CONTROL OF CARBON DIOXIDE

the dynamic processes that determine climate have not attained a stable equilibrium. One of the more carefully documented examples is the global mean temperature, which over the last 100 years has shown a range of variation of five-year averages of about 0.6°C (see figure 8.6). The disputes about the sources of such variations are reminiscent of business cycle theory: theories encompass everything from sunspots to quasi-periodic oscillations to the existence of many locally stable but globally unstable equilibria.

At what point is there likely to be a significant[2] effect of man's activities on the climate? Many climatologists feel that the changes witnessed in the last century—the 0.6°C range— have led to major, albeit not catastrophic, results. It should be stressed that the changes in temperature are rather trivial. Mean temperature changes of this size are not economically significant. Rather, the critical variables are degree-days, precipitation, and snow cover, and these tend to vary *much* more than global mean temperature. Examples of high amplification are changes in precipitation and changes in the latitude of monsoons with changes in temperature. [See Machta and Telegados (1974) or Schneider (1976)]

If we define a significant change arbitrarily at a 0.5°C change in global mean temperature, for carbon dioxide such a change would come with about a 20 percent increase in atmospheric concentrations, and it appears that carbon dioxide will be the first man-made emission to affect climate on a global scale.[3]

A.2. *Climatic Effects of Carbon Dioxide*

Combustion of fossil fuels leads to emissions of carbon dioxide into the atmosphere. The emissions slowly distribute themselves by natural processes into the oceans, into the

2. "Significant" in the economic, not the statistical sense.

3. The estimates of carbon dioxide concentration are from Baes et al. (1976) and are consistent with my estimates in section F below. The temperature response is from Manabe and Wetherald (1967). The more recent estimate in Manabe and Wetherald (1975), which is 50 percent larger than their earlier estimate, is not used because this includes the full ice-albedo feedback (including land-based ice), which can hardly be expected within 50 years.

biosphere, and, eventually, into fossils. Although this exact process is not completely understood, it is clear that the residence time of carbon dioxide in the atmosphere is extremely long and that at the present approximately half the industrial carbon dioxide remains in the atmosphere. The ultimate distribution of carbon dioxide between the atmosphere and the other sinks is not known, but estimates of the man-made or industrial carbon dioxide asymptotically remaining in the atmosphere range between about 2 and 50 percent.[4]

It is generally thought that there are two important effects of the atmospheric buildup of carbon dioxide. First, there may be a beneficial effect of increased concentration on agriculture, since higher concentrations lead to higher rates of photosynthesis.[5] The second effect is on the climate through the "greenhouse" effect. Because of the selective absorption of radiation, the increased carbon dioxide concentration leads to an increase in the surface temperature of the earth. A recent study by Schneider (1976), listing recent studies, gives a range of estimates from 0.7°C to 9.6°C. The Manabe-Wetherald (1975) estimate is the most complete for long-run purposes in that it is a three-dimension general circulation model, with the feedback effects between temperature and snow-ice-albedo. When discussing long-run effects, I will therefore consider primarily Manabe-Wetherald (1975) in the present discussion, although for the short-run (up to 100 years), the early Manabe-Wetherald (1967) is more appropriate in that it includes no ice-albedo feedback.

Figure 8.1 shows estimates in Manabe-Wetherald (1975) of the long-run effect of CO_2 doubling on surface temperature by latitude. The effect on surface temperature is generally around 2°C up to about 40° latitude (roughly New York), then increases dramatically, to 4.5 degrees at 60°, up to over 10°C in the polar regions. Judging by the difference of temperatures in two standard runs [Manabe-Wetherald (1975), p. 15] temperature differences above 1°C are statistically signi-

---

4. See Matthews et al. (1971), Machta (1972), Keeling (1973b), Baes et al. (1976), MacIntyre (1970), GARP (1975).

5. For a further discussion, see section C.2.

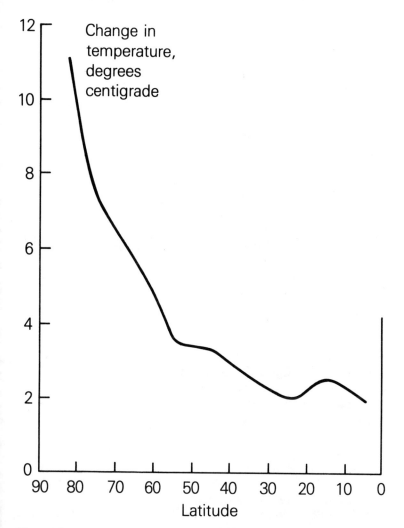

Figure 8.1.
Estimated effect of doubling of atmospheric carbon dioxide
on surface temperatures, by latitude. From Manabe and
Wetherald (1975).

ficant. The calculated effect on precipitation is to increase precipitation in latitudes above 40°, and is mixed below 40°. Precipitation is predicted to decrease in the 30–40° belt and in the desert regions (10–20°).

Beyond the results on temperature, which are generally agreed on to an order of magnitude, several authors paint somber pictures of the climatic response to the warming trend. One of the more detailed pictures is that of the Soviet climatologist M. I. Budyko [see (1972) and (1974)]. His studies indicate that one of the most critical points in the earth's climatic equilibrium is the floating Arctic ice pack. This varies from 2 to 3 meters in thickness and is quite sensitive to minor temperature fluctuations. According to Budyko, a summer temperature increase of 4°C would lead to the melting of the floating Arctic ice pack within a decade. From figure 8.1 we see that the temperature increase from doubling CO_2 is well beyond 4°C. If Budyko is correct, it is highly probable then that the permanent Arctic ice pack will disappear well before the time CO_2 doubling occurs.[6]

By itself, an open Arctic ocean would lead to rather dramatic changes in the climate of the Northern hemisphere. A recent simulation experiment by Newson (1975) attempts to resolve more finely the effects. The experiment suggests that it is possible that even though the climate as a whole is warmed, a cooling of continental climates may occur because of weakened westerlies; Newson's study predicts the continental United States will *cool* by 8°C.

It is crucial to separate the floating Arctic ice from the land-based ice. Melting of sea-ice has no effect on sea level, while a rapid melting of the massive ice caps of Greenland and Antarctica would be a major catastrophe, for the ice caps contain enough water to raise the sea level by 90 meters (300 feet). Past evidence indicates, however, that any melting will proceed

6. It should be noted that studies of Manabe and Wetherald (1975) do not indicate melting of polar ice in the CO_2 doubling, but these simulations contain no seasonal features and misspecify certain crucial geological features (for example they assume that above 70° latitude the earth is land rather than sea).

extremely slowly. In previous warm periods, the period over which glacial retreat occurred was in the order of 5,000 years (see below).

B. *Strategies for Control*

The outcome just described is the effect of an *uncontrolled* economy-climate system, one in which the economy, the energy system, the emissions of carbon dioxide, and the climatic response evolve simply on the basis of economic forces. The problem is a classical example of economic externality. An externality arises when economic agents do not pay for the entire social cost of their activities. Thus, when a steel mill spews out soot that blackens the neighborhood, the owners of the mill pay for the labor and capital they use but not for the higher laundry or doctor bills caused by the dirty air. From a private point of view, clean air has a zero price to the steel factory, and, being so cheap, it is natural that cost-conscious managers will substitute cheap air for expensive labor and capital. From a social point of view the result is too much steel and dirty air, too little pollution abatement.

In analyzing the effect of man's impact on climate we are faced with a pure example of an externality. When an individual or firm burns gasoline in a car, or oil in a furnace, he pays for the capital equipment in the furnace and for the fuels. He pays nothing for his carbon dioxide emissions or the effect of his activities on the climate. Even if he is an altruist, he would have to recognize that his contribution to solving the long-run climate problem is negligible.

The control problem for carbon dioxide thus involves two aspects. On a scientific and aggregate level, the feasibility of controls and control techniques must be explored. But there must also be a way of decentralizing the control (of internalizing the externality) so that individual producers and consumers have proper incentives to implement the control strategy on an individual level.

There are several general approaches to the problem of keeping atmospheric concentrations to a reasonable level. At the top of the list (in likelihood if not desirability) is the approach of doing nothing. This consists of simply letting the market forces provide the solution, with the price of climatic change and disruption implicitly set at zero. The other strategies are active in that they attempt to reduce atmospheric concentrations to a "tolerable" level. In the category of active strategies, there are three possibilities. The first, which is the route chosen in the present study, is to reduce emissions of carbon dioxide. This takes place by substituting non-carbon-based fuels for carbon-based fuels.

The second strategy is to offset the effects of emissions of carbon dioxide. This can take the form of introducing the carbon into places where its climatic effect is nullified or delayed or of using counteracting forces to offset the effects. For example, if CO_2 is compressed and pumped into the oceans at a depth of at least 2,000 meters, it would be at a specific gravity heavier than water and would therefore tend to remain at great depths until molecular or eddy diffusion raised it to the mixed layer—probably only after thousands of years. Other suggestions have been to introduce stratospheric dust to cool the earth, to change the earth's albedo by putting gauze over the arctic, or to paint roads or roofs white.

A third approach would be to use natural or industrial processes to clean out the carbon dioxide from the atmosphere ex post. This approach would rely on the possibility that removing the atmospheric carbon dioxide is cheaper than refraining from putting the carbon in the atmosphere in the first place. Possibilities here include simply growing trees and locking the carbon in the trees or removing the carbon from the air by an industrial process.

To avoid the image of science fiction, I have initially limited control strategies to those clearly feasible—reductions in demand and substitution in supply.

The second problem of controlling carbon dioxide is implementation on a decentralized level.

Once some notion about an efficient path has been obtained, there must be a way of assuring that the millions of economic actors have incentives to reduce emissions. In the real world, the policy can take the form either of taxing carbon emissions or of physical controls (such as rationing). In an efficient solution, the two are interchangeable in principle; in practice, the use of taxes is much simpler because the taxes tend to be much more uniform than the quantities. I therefore will concentrate on "carbon taxes" as a way of implementing the global policy on a decentralized, individual level.

C. Dynamics of the Carbon Dioxide Cycle
C.1. Sources of Carbon Dioxide

Keeling has recently described quite carefully the origins of man-made carbon dioxide.[7] Approximately 98 percent of industrial carbon dioxide originates in the energy sector, although of this about 5 percent ends up in nonenergy uses (in asphalt, bitumen, lubricants, etc.). The other 2 percent of the industrial source is cement production. Table 8.1 gives the conversion factors for deriving the emissions of carbon dioxide from the consumption of fossil fuels, as well as the assumed conversion factors for nonfossil technologies.

Table 8.1.
Emission Factors for Carbon Dioxide

	Carbon fraction in fuel by weight	Fraction of fuel oxidized	Conversion factor (tons carbon per ton fuel)	Carbon content (10^9 tons carbon per 10^{15} btu)
Coal and lignite	0.70	0.99	0.693	0.0279
Crude petroleum	0.84	0.915	0.769	0.0239
Natural gas	n.a.	0.97	n.a.	0.0144
Electrolytic hydrogen	0	n.a.	0	0
Nuclear energy	0	n.a.	0	0
Solar	0	n.a.	0	0

SOURCE: For fossil fuels, from Keeling (1973a), pp. 191, 180, 181, 178. The conversion factors (from Keeling) are 12,400 btu/lb for coal and lignite, 19,000 btu/lb for petroleum, and 1,030 btu per cu ft for natural gas.

n.a. = not applicable.

7. Keeling (1973a).

C.2. *Diffusion of Atmospheric Carbon Dioxide*

Once carbon dioxide enters the atmosphere, the process of diffusion into the ultimate reservoirs begins. Compared with most atmospheric pollutants, this process is extremely slow. Thus according to Keeling (1973a), man's activities have added 18 percent to the atmospheric carbon dioxide over the period 1860 to 1969; of the 18 percent approximately 10 percent, or 65 percent of the total added, remains in the atmosphere. A more recent study by Baes et al. (1976) suggests that 55 percent remains airborne.

In the modeling, I will use a seven-layer well-mixed box system, following the work of Keeling and Machta. Figure 8.2 shows the major reservoirs and flows in the system. The major assumption is that all reservoirs are well-mixed, and that the transfer (or diffusion) between boxes follows a first-order exchange process. Under this set of assumptions, the system of equations is thereby linear and easily incorporated into a mathematical programming framework.

The strata in the model are two atmospheric strata (stratosphere and troposphere); two ocean layers (mixed ocean—down to 60 meters—and deep layer); and three biospheres (short-term land biosphere, long-term land biosphere, and marine biosphere). In estimating the transfer coefficients in figure 8.2, most are from extraneous information rather than direct estimates. The two coefficients relating to the transfer between the troposphere and the mixed layer, however, are estimated empirically by Machta using residence times from bomb-C^{14}.

It is generally agreed that the concentrations in the atmosphere and the mixed layer of the oceans comprise about 60 percent of the industrial carbon dioxide. The simple model used here also predicts approximately that breakdown. From what is generally known about the age of the deep ocean, not much more than 5 percent of the industrial carbon dioxide would be in the deep oceans at present, but assigning the remaining 35 percent to an increase in the biosphere seems implausible. As the area of greatest controversy and the most likely reservoir was the deep oceans, I increase the transfer from mixed to deep layer by changing the average age (or residence time) from 1,700 to 800 years—a figure well within

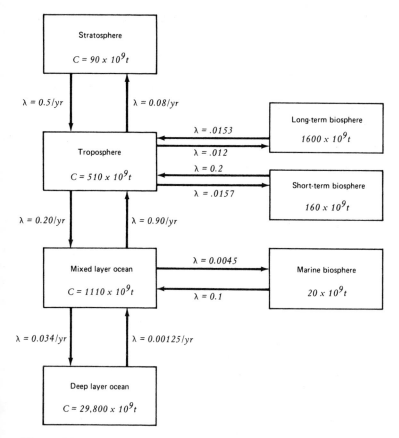

Figure 8.2.
The marginal first order transfer process between the seven
reservoirs of carbon dioxide. The λ are the transfer coefficients,
indicating what fraction of the mass of one reservoir is trans-
ferred to the second reservoir per year. The figures give the
estimated amount of carbon (in metric tonnes) in each reservoir
in preindustrial times.

SOURCES: Machta (1972) as updated in Machta and Telegados (1974), p. 696.
Exceptions are the biomass estimates, from Baes et al. (1976), and the transfer
coefficients in the oceans, discussed in the text. Note that the biosphere
transfer coefficients shown are equal to the biosphere-uptake elasticity (η)
times the average or observed transfer coefficients, and that it is assumed
that the buffer coefficient (b) is 10.

the range of estimates. This correction seems to be consistent with recent data on aging [see Stuiver and Broecker (1975)], as well as the calculations of Oeschger et al. (1975).

The technical operation of the model can be easily shown. Let d_{ij} be the transfer coefficient per year from reservoir i to reservoir j; let the one-year transfer matrix $[d_{ij}]$ be represented by D. Note that D is a Markov matrix, so $\sum_{j=1}^{7} d_{ij} = 1$. Further, let the mass of a given reservoir in year t be denoted by $M_i(t)$, $i = 1, \ldots, 7$; with the column vector $M(t)$.

Our basic diffusion equation is that:

$$M_i(t) = \sum_{j=1}^{7} d_{ji} M_j(t - 1), i = 1, \ldots, 7,$$

or in matrix form

$$M(t) = D'M(t - 1),$$

where D' is the transpose of D.

Table 8.2 shows the one year and the 100 year transfer matrix. These indicate how a one-shot injection into a given stratum is distributed over the indicated period.

An examination of the more detailed set of distribution coefficients for atmospheric emissions [in Nordhaus (1977), not shown] shows how the emissions are distributed for continuous emissions for the given time period. The estimates for the fraction of the CO_2 remaining in the atmosphere are slightly higher than in most other models for the short run—with 77 percent remaining in the atmosphere after one year; or 62 percent after twenty years. While this figure exceeds some estimates [see Machta (1972), PSAC (1965), Keeling (1973a)], it should be noted that these are *marginal* residences for a twenty-five year period whereas other figures cited refer to the average residence time of *all* man-made carbon dioxide. The asymptotic fraction of the total carbon dioxide remaining in the atmosphere is 15 percent, a figure well below the usual assumption in simple calculations.

D. *Limits on Carbon Dioxide Concentrations*

Up to now, there has been no serious thought of the level of standard on carbon dioxide. As a first approximation, it

Table 8.2.

One Year and 100 Year Distribution Matrixes

One Year Distribution Matrix, $b = 10, \eta = 0.25$

	T	S	M	D	SB	LB	MB
T	0.688	0.088	0.196	—	0.016	0.012	—
S	0.500	0.500	—	—	—	—	—
M	0.900	—	0.061	0.034	—	—	0.005
D	—	—	0.00125	0.99875	—	—	—
SB	0.200	—	—	—	0.800	—	—
LB	0.015	—	—	—	—	0.985	—
MB	—	—	0.100	—	—	—	0.900

100 Year Distribution Matrix, $b = 10, \eta = 0.25$

	T	S	M	D	SB	LB	MB
T	0.301	0.055	0.065	0.289	0.027	0.262	0.001
S	0.303	0.055	0.066	0.285	0.027	0.262	0.001
M	0.300	0.053	0.064	0.310	0.025	0.254	0.001
D	0.050	0.009	0.011	0.905	0.004	0.022	0.000
SB	0.314	0.056	0.067	0.280	0.027	0.262	0.001
LB	0.310	0.054	0.065	0.157	0.026	0.390	0.001
MB	0.314	0.056	0.067	0.291	0.027	0.252	0.001

NOTES ON MATRIX: The distribution matrix is a probability matrix whose rows each sum to one. The entries indicate the fraction of the mass of that basin on the left-hand column that flows per unit time period to the basin on the top row. The basins are denoted as follows:

T = Troposphere

S = Stratosphere

M = Mixed layer of the oceans (0 to 60 meters deep)

D = Deep layer of the oceans (deeper than 60 meters)

SB = Short-term biosphere

LB = Long-term biosphere

MB = Marine biosphere

b = buffering coefficient for oceans

η = elasticity of carbon dioxide uptake in biosphere with respect to carbon dioxide concentrations.

seems reasonable to argue that the climatic effects of carbon dioxide should be kept within the normal range of long-term climatic variation. According to most sources, the range of variation between distinct climatic regimes is in the order of $\pm 5°C$; at the present time the global climate is at the high end of this range. If there were global temperatures more than 2 or 3°C above the current average temperature, this would take the climate outside the range of observations that have been made over the last several hundred thousand years. Within a stable climatic regime, such as the current interglacial, a range of variation of 2°C is the normal variation. In these standards, we must ignore any background increase or decrease in atmospheric carbon dioxide, as well as the effects of particulates and other contaminants. These are ignored because scientific knowledge about their magnitude and effects is inadequate.

Thus, I assume that doubling of atmospheric concentration of carbon dioxide is a reasonable upper limit. I will also test the sensitivity of my results to limits by imposing limits of 50 percent and 200 percent increase. *It must be emphasized that the standards proposed here are deeply unsatisfactory, from both an empirical point of view and a theoretical point of view.* I am not certain that I have even judged the *direction* of the desired movement in carbon dioxide correctly, to say nothing of the absolute levels.

E. *The Carbon Dioxide-Energy Model*

The energy model used for the investigation is fully described in chapter 4; it is a linear programming model designed to simulate the functioning of a competitive market for energy products.

To add carbon dioxide to the model, we need to introduce the three factors discussed in the last section: emissions, diffusion, and standards. First, let $\gamma(\ell\ell,i)$ be the emissions per unit activity x_{it} into stratum $\ell\ell$ (in 10^9 tons carbon per 10^{15} btu). Then total emissions into stratum $\ell\ell$ in a given period, $E(\ell\ell,t)$ are

$$E(\ell\ell,t) = \sum_{i=1}^{n} \gamma(\ell\ell,i)x_{it}, \ell\ell = 1, \ldots, L, t = 1, \ldots, T.$$

Next denote $M(\ell\ell,t)$ as the total mass of CO_2 (in 10^9 tons C) in a given stratum and $d(i,j)$ as the transition probability of moving from stratum i to stratum j. From the basic diffusion equations we have

$$M(\ell\ell,t) = \sum_{i=1}^{L} d(i,\ell\ell) M(i,t-1) + E(\ell\ell,t),$$
$$\ell\ell = 1, \ldots, L, t = 1, \ldots, T.$$

Finally, I impose standards on the energy sector that the total mass in a given stratum should not exceed $St(\ell\ell)$:

$$M(\ell\ell,t) \leqq St(\ell\ell), t = 1, \ldots, T.$$

To implement the controls, I add the above equations to our original model discussed in chapter 4. It should be noted that the optimization framework makes computation of a single run relatively expensive, precluding extensive experimentation and sensitivity analysis.

F. *Results of the Standards Model*

This section presents the results of the runs with the "standards model" outlined in the last section. Recall that there are four different runs; they differ only in the standards imposed on the concentration of carbon dioxide.

F.1. *The Question of Feasibility*

The first question to investigate is whether the standards paths are feasible. The question of feasibility rests on the existence of activities that meet the demand constraints with relatively low levels of carbon dioxide emissions. In reality, any nonfossil fuel energy source (fission, fusion, solar, or geothermal) will be an option for meeting the carbon dioxide constraint since the nonfossil fuels have no significant carbon dioxide emissions. In the program discussed above, both solar and nuclear fission are considered as an alternative to fossil fuels, but the results would be identical for any of the other nonfossil fuels (fusion, geothermal) *with the same cost structure.*

In the program outlined above, it would be possible to set arbitrarily low carbon dioxide standards because the energy system can adapt to these by simply shifting the mix from fossil to nuclear fuels. It should be noted, however, that the model used here overemphasizes the degree of malleability of the system by ignoring the fixity of historically built capital equipment as well as overestimating the speed of reactions. To be realistic, it is probable that it would take at least 25 years to phase out carbon-based fuels even with a crash effort, so this places an outside limit on the feasibility of carbon dioxide limitation. Aside from this lag, and assuming the technological relations are correctly specified, however, there are no significant *technical* problems of severely limiting carbon dioxide emissions.

F.2. *Comparison of Uncontrolled and Controlled Programs: Quantities*

The next question concerns the comparison of the uncontrolled path and the controlled paths. In the program dis-

Table 8.3.
Industrial Carbon Dioxide Emissions and Concentration Predicted from Model

Industrial carbon dioxide emission rate	Pre-industrial	Actual 1974	Projected					
			1980	2000	2020	2040	2100	2160
(10⁹ tons carbon/yr)								
1. Uncontrolled			6.9	10.7	18.3	40.1	64.0	0.0
2. 200% increase	0.0	5.0	6.9	10.7	18.3	38.1	36.6	14.7
3. 100% increase			6.9	10.7	16.6	16.1	4.9	3.7
4. 50% increase			6.9	8.9	4.0	2.4	1.6	1.3
Total carbon dioxide concentration in atmosphere			1990	2010	2030	2050	2070	2170
(10⁹ tons carbon)								
1. Uncontrolled			768	896	1115	1615	3137	2355
2. 200% increase	616	702	768	896	1115	1586	1951	1953
3. 100% increase			768	896	1088	1244	1239	1237
4. 50% increase			768	869	883	879	871	868

NOTE: CO_2 in the atmosphere in 1970 is distributed over time according to the distribution model and then added to the calculated amount. This procedure introduces a minor inaccuracy in the optimization procedure.

cussed above, I have divided the period into 10 periods, each
with 20 years. The most important question is the timing of
the limitations on carbon dioxide emissions. Table 8.3 and
figures 8.3 and 8.4 show the paths of emissions and concentrations for carbon dioxide in the atmosphere for each of the four
paths.

The first point to note is that the uncontrolled path does
lead to significant changes in the level of atmospheric carbon
dioxide. According to the projection of the model, atmospheric
concentrations in the uncontrolled path rise by a factor of five

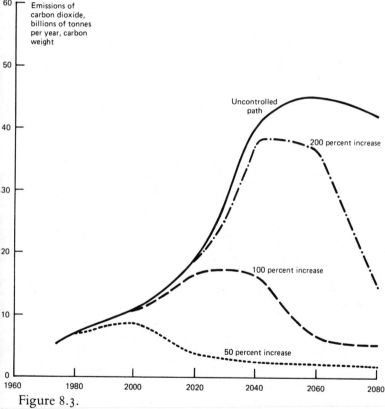

Figure 8.3.
Calculated emissions of carbon dioxide along alternative paths,
1960–2080, with actual emissions for period up to 1974.
Figures in billions of metric tonnes, carbon weight.

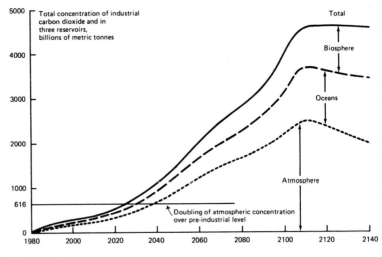

Figure 8.4.
Distribution of industrial carbon dioxide over time by reservoir, uncontrolled path (billions of metric tonnes).

(3137/616) over the entire period. This increase is far above our arbitrary limit of a doubling of the carbon dioxide concentration. Put differently, it appears that *if serious problems are likely to occur when the level of carbon dioxide has doubled or more, then the uncontrolled path appears to be heading for the danger zone.* It appears that the doubling will come around 2040.

Table 8.4 compares the calculated path with other estimates of emissions and concentration. The emissions agree very well with concentrations and actual figures. Projections for the future are also shown.

The second important point, and perhaps the most surprising one, is that the optimal path does not differ from the uncontrolled path for the first periods (that is to say the periods from 1970 to 1990) and that abatement measures become necessary only in the second period (1990 to 2010) for the most stringent controls and the third period (2010 to 2030) for the other programs. According to the cost schedules assumed in the model, it does not pay to curtail carbon dioxide emissions until nearly the time when the limit is reached; and for the

Table 8.4.
Comparison of Uncontrolled Model Projections with Observed Values, 1974, and Other Projections, 2000

| | 1974 | | 2000 | | |
	Actual	Calculated from model	Calculated from model	Estimated by: Machta	Baes et al.
Atmospheric Concentration					
In 10^9 tons carbon	702.	702.	832.	846.	702. to 862.
In parts per million	321.	321.	381.	387.	321. to 394.
Emission					
In 10^9 tons carbon	5.0	4.9	10.7	11.4	

SOURCES: Actual from Baes et al. (1976). Calculated for 1974 use actual values for 1970 and interpolate geometrically. Figures for Machta from Machta and Telegados (1974), p. 695; for Baes et al. (1976), p. 39.

three cases examined this time comes in the 1990–2010 period or the 2010–30 period. This point is important, for it implies that there is still some time to continue research and to consider plans for implementation of carbon dioxide control if it is deemed necessary.

The program calculates, but I have not shown, the effect of the constraints on demand. Recall that demand is sensitive to price, so that it is possible that demand will be curtailed in order to meet the carbon constraints. At first blush, it is plausible to argue that since carbon emissions must be reduced by 80 percent from the uncontrolled path, demand must also be reduced by 80 percent. In fact, this naive view would be almost completely wrong: almost no changes in the demand pattern occur, and almost all the reaction comes about as a result of supply side adjustments. Put differently, *the efficient way to restrict emissions is to change the composition of production away from carbon-based fuels rather than to reduce consumption.* Figure 8.5 shows for the United States the effect of the carbon dioxide controls on gross energy inputs (usually called "energy consumption"). The striking result is that very little change in end

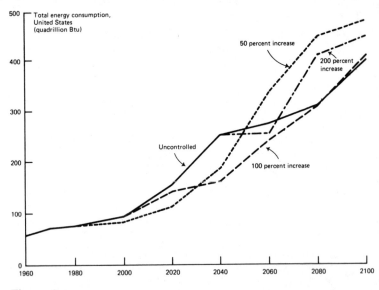

Figure 8.5.
Total energy consumption (gross energy inputs), United States, for alternative control programs. Figures for 1960–75 are actual, those for 1980–2100 are calculated in alternative control programs.

use or energy inputs is required to meet the carbon dioxide constraints.

F.3. *Effects of Control Programs on Temperature and Sea Level*

Figure 8.6 shows the estimated effect of different control programs on global mean temperature. The relationship assumed in this figure is that temperature is a function of the logarithm of CO_2 concentration, and that a doubling of CO_2 leads to an increase of 2°C. To estimate the response of the oceans is a most hazardous exercise. Long-run equilibrium models such as Manabe-Wetherald (1975) mask much of the dynamics of climate change. At present, there are no generally accepted models that show the dynamics of the response of land-borne ice to future temperature changes. As a result it is difficult to judge what fraction of the predicted climatic effects

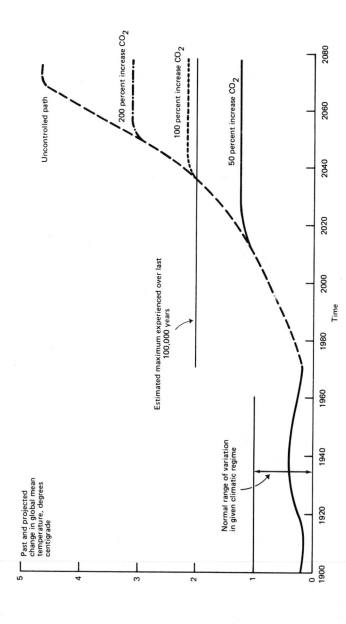

Figure 8.6.
Past and projected change in global mean temperature, relative to 1880–84 mean. Solid curve up to 1970 is actual temperature. Curves from 1970 on are projections using 1970 as a base and adding the estimated increase due to carbon dioxide.

will occur within the next 150 years. Given the importance of the response of the oceans to temperature, I attempted in Nordhaus (1976) to obtain very crude estimates of the relationship. The data for the last 15,000 years indicates that a 1°C increase in temperature would lead to an asymptotic increase in the level of the oceans of between 5 and 10 meters. On the other hand, the response appears to be extremely slow, with the time period for one-half the rise being in order of 1,500 to 3,000 years. I thus estimate that the order of magnitude rise of oceans is 2.4 ($\pm$1.0) millimeters per year per degree C increase. Note that currently the oceans are rising about 1 mm per year.

Using these estimates, I can make a rough projection of the rise in sea level over the next century or so. Along the

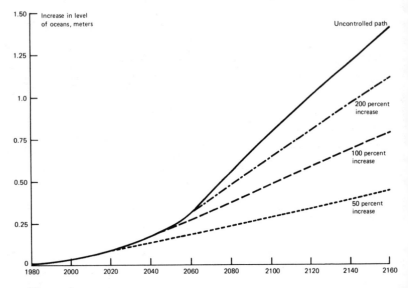

Figure 8.7.
Estimates of the effect of temperature increase on the level of oceans for alternative paths of carbon dioxide concentrations. All calculations assume no rise in coastlines with no change in carbon dioxide concentrations. Estimate of temperature increase is shown in figure 8.6. Effect on sea level is assumed to be 2.4 millimeters per degree C per year [see appendix A of Nordhaus (1976) for derivation].

estimated uncontrolled path, the cumulative rise predicted on the two assumptions would be about 0.1 meters by 2000, 0.3 meters by 2050, and 0.8 meters by 2100. Figure 8.7 uses this estimate along with the estimates of global temperature increase in figure 8.6 to indicate an order of magnitude estimate of the effect of rising CO_2 concentrations on the level of the oceans. Per se, changes of these magnitudes and at this speed are not catastrophic, although they would eventually cause hardship in low-lying areas. The major danger is not the more-or-less predictable rise that would accompany the warming trend, but less predictable events such as Antarctic ice surges.

F.4. *Prices and Costs*

In an optimization framework, as in an economy, constraints have their costs in terms of the objectives of the optimization. Recall that the control program takes the form of imposing contraints (or upper bounds) on the level of atmospheric concentrations. Associated with each of the constraints is a dual variable—a "shadow price"—that in the optimal solution calculates the incremental amount that the constraint costs in terms of the objective function. Put differently, the shadow price indicates how much the objective function would increase if the constraint were relaxed one unit.

Table 8.5 gives the shadow prices for carbon emissions for the four programs during the ten periods. The uncontrolled program has shadow prices equal to zero, indicating that the constraint is not binding. In controlled programs, the prices per ton start very low (between $0.01 and $0.15 per ton carbon) and rise to a very high level of between $130 a ton (1975 prices) by the end of the next century. These should be compared with the prices of carbon-based fuels, which are around $25 a ton (carbon weight) of coal, $100 a ton (carbon weight) for petroleum, and $200 a ton (carbon weight) for natural gas. Roughly speaking, the shadow price only becomes significant in the second period for the most stringent path (path IV) and in the third and fourth period for the medium and permissive paths, III and II respectively.

The shadow prices on carbon dioxide play a leading role in the drama. Not only do they show the cost of a given con-

Table 8.5.

Shadow Prices on Carbon Dioxide Emissions (1975 dollars per ton carbon)

	Program			
	I Uncontrolled	II 200% Increase	III 100% Increase	IV 50% Increase
1980	0.00	0.00	0.14	1.65
2000	0.00	0.07	1.02	12.90
2020	0.00	0.52	8.04	109.00
2040	0.00	4.07	67.90	123.60
2060	0.00	34.47	94.40	200.00
2080	0.00	42.00	94.40	200.00
2100	0.00	42.04	87.20	198.20
2120	0.00	41.91	87.10	198.50
2140	0.00	42.92	86.90	188.40

straint for the world economy as a whole; they also are the best mechanism for decentralizing the control strategy. It is essential for implementing a control strategy that the shadow prices on carbon emissions actually get built into prices that firms and consumers face. Without such an "internalization" of the climatic externality, it can hardly be expected that emissions will be reduced: by contrast, when coal prices double as a result of an emissions tax, we can expect substitution away from coal and its derivatives.

We may also ask what the effect of the carbon dioxide control program is on energy prices in general. These effects fall into two general categories: Effects on factor prices—in particular royalties on scarce energy resources; and effects on product prices. Note that the major impact [shown in Nordhaus (1976)] is on factor prices rather than product prices. For example, in the most stringent case, the shadow prices of petroleum and gas shadow prices fall considerably for the abundant non-U.S. resources, while coal and oil shale royalties *fall to zero*. By contrast, uranium royalties *rise* by an insignificant amount (about 0.1 percent) from the uncontrolled to the most stringent program since they are a non-carbon-based fuel.

For final product prices (including distribution, taxes, and retail markups where appropriate) the uncontrolled path shows essentially no price increase from the 1980 levels for about 40 years, then an upturn in prices with the exhaustion of fossil fuels and the gradual penetration of nonfossil fuels late in the next century. The two less stringent control programs look very similar, with only very minor increases in prices (less than 5 percent higher than the uncontrolled path). The most stringent control path, however, shows a much more rapid increase in prices over the next fifty years; it is economically equivalent to having less fossil fuel resources in that the stringent control program drives up prices sufficiently to ensure more rapid penetration of nonfossil fuels.

A final question regarding shadow prices may appear bizarre: What are the shadow prices on emissions by environmental *stratum*? These refer to the shadow prices in the different regions of the earth (atmosphere, mixed ocean, deep ocean, etc.). Table 8.6 shows the shadow prices for each of the seven strata for the middle control strategy, again in terms of prices per ton of carbon. These indicate the cost that would be incurred by an increase of one ton of the mass in a given stratum. Thus the price for carbon in the atmosphere in 2020 would be $9, while in the long-term biosphere it would be $3.

The important point about table 8.6 is that there are only three economically interesting strata: the deep ocean, the long-term biosphere, and the rest of the strata. And the most interesting conclusion is that the cost of putting carbon into the deep ocean is only about one-thirtieth of the cost of putting it into the atmosphere. The reason for this anomaly is that by the time carbon is put into the deep ocean it is locked up there for about 1,000 years. The price in the long-term biosphere is also significantly below, eventually about one-third, the price in the other strata.

The implication of this finding about the shadow prices in different strata is of great importance for control programs. It says that on the margin, and taking 2020 as an example, it

Table 8.6.
*Shadow Prices on Carbon Dioxide Concentrations by Stratum,
Control Program III (dollars per ton carbon, 1975 prices)*

	Period Centered on:			
	1980	2000	2020	2100
Troposphere	0.14	1.09	8.75	76.90
Stratosphere	0.14	1.09	8.75	76.90
Mixed layer ocean	0.14	1.06	8.53	73.50
Deep layer ocean	0.01	0.10	0.29	2.30
Short-term land biosphere	0.14	1.14	9.10	78.70
Long-term land biosphere	0.11	0.66	3.15	26.10
Marine biosphere	0.14	1.08	8.69	75.00

would be efficient to take emissions from the atmosphere and pump them into the deep oceans if this could be done for less than $8 per ton. Similarly, if we could simply remove the carbon and put it into trees, which would decay, gradually adding the carbon back into the atmosphere, this would be worth a subsidy of up to $6.50 per ton. These results can be used to evaluate processes to short-circuit the distribution of carbon dioxide by placing it in the deep ocean or in trees. Given some preliminary estimates of the costs of these processes, it appears that they merit considerable attention.

We can also ask what the carbon dioxide constraints are costing in toto. Whereas the shadow prices give the cost on the margin, the overall cost can be evaluated by examining the attained value of the objective function. The control of carbon dioxide is not free—the three control programs have discounted costs of $4, $87, and $540 billion in 1975 prices. On the other hand, the cost as a fraction of world GNP is likely to be insignificant, less than 0.5 percent in the most stringent case.

REFERENCES

Adelman, M. A. (1972). *The World Petroleum Market*. Johns Hopkins Press for Resources for the Future, Baltimore, Md.

Ahrens, L. H. (1954). The Lognormal Distribution of the Elements. *Geochimica et Cosmochimica Acta* 5:49–73.

Allais, Maurice (1957). Method of Appraising Economic Prospects of Mining Exploration over Large Territories. *Management Science* 3:285–347.

Baes, C. F., M. E. Goeller, J. S. Olson, and R. M. Rotty (1976). *The Global Carbon Dioxide Problem*. Oak Ridge National Laboratory, Oak Ridge, Tenn.

Balassa, Bela (1964). Purchasing Power Parity. *Journal of Political Economy* 73:584–96.

Barouch, Eytan, and Gordon Kaufman (1976). *Oil and Gas Discovery Modeled as Sampling Proportional to Random Size*. Working Paper. Alfred P. Sloan School of Management, MIT, Cambridge, Mass.

Baumol, William J. (1971). *Economic Theory and Operations Analysis*, 3rd ed. Prentice-Hall, Englewood Cliffs, N.J.

Beller, M. (ed.). (1975). *Sourcebook for Energy Assessment*. Brookhaven National Laboratory, BNL-50483, Upton, N.Y.

Blitzer, C., A. Meeraus, and A. Stoutjesdijle (1975). A Dynamic Model of OPEC Trade and Production. *Journal of Development Economics* 2:319–35.

Brinck, Johan W. (1967). Note on the Distribution and Predictability of Mineral Resources. Euratom Report: EUR 3461e, Brussels, Belgium.

Budyko, M. (1972). The Future Climate. *Transactions of American Geophysical Union* 53:868–74.

——— (1974). *Climate and Life*, Chaps. 3–5, 9. Academic Press, New York.

Coulomb, R. (1959). Contribution à la Géochémie de l'uranium dans les granites intrusifs. Rapport, C.E.A. no. 1173.

Cremer, J., and M. L. Weitzman (1976). OPEC and the Monopoly Price of World Oil. *European Economic Review* 8:155–64.

Energy Facts (1975). House Subcommittee on Energy Research, Development and Demonstration. *Energy Facts II.* Government Printing Office, Washington, D.C.

Ezzati, Ali (1976). Future OPEC Price and Production Strategies as Affected by Its Capacity to Absorb Oil Revenues. *European Economic Review* 8:107–38.

FEA (1976). Federal Energy Administration. *National Energy Outlook.* Government Printing Office, Washington, D.C.

Fettweis, Gunter B. (1975). Contributions to the Assessment of World Coal Resources or Coal is Not so Abundant. Unpublished mimeo. IIASA, Laxenburg, Austria.

Fisher, J. C., and R. M. Pry (1970). A Simple Substitution Model of Technological Change. General Electric Report 70-C-215, Schenectady, New York.

FTC (1974). Federal Trade Commission. *Concentration Levels and Trends in the Energy Sector of the U.S.* Government Printing Office, Washington, D.C.

GARP (1975). The Physical Basis of Climate and Climatic Modelling. World Meteorological Organization, Garp Publications Series, no. 16, Geneva, Switzerland.

Grenon, Michel (1975). Studies on Energy Resources in the IIASA Energy Project. IIASA Research Report, Laxenburg, Austria.

Hendricks, T. A. (1965). Resources of Oil, Gas, and Natural Gas Liquids in the United States and the World. U.S. Geological Survey Circular 552, Government Printing Office, Washington, D.C.

Hnyilicza, E., and R. S. Pindyck (1976). Pricing Policies for a Two Part Exhaustible Resource Cartel: The Case of OPEC. *European Economic Review* 8:139–54.

Hoffman, Kenneth (1972). The U.S. Energy System—A Unified Planning Framework. Unpublished PhD Dissertation, Brooklyn Polytechnical Institute, Brooklyn, N.Y.

Hotelling, Harold (1931). The Economics of Exhaustible Resources. *Journal of Political Economy* 39:137–75.

Houthakker, H. S. (1976). The World Price of Oil: A Medium Term Analysis. American Enterprise Institute, Washington, D.C.

Hubbert, M. K. (1969). Energy Resources. In *Resources and Man*, National Research Council Committee on Resources and Man. Freeman, San Francisco, Calif.

Kalymon, B. A. (1975). Economic Incentives in OPEC Oil Pricing Policy. *Journal of Development Economics* 2:337–62.

Keeling, Charles D. (1973a). Industrial Production of Carbon Dioxide from Fossil Fuels and Limestone. *Tellus*: 25:174–98.

———— (1973b). The Carbon Dioxide Cycle. In N. Rasool (ed.). *Chemistry of the Lower Atmosphere.* Plenum Press, New York, N.Y. Pp. 251–329.

Kellogg, W. W., and S. H. Schneider (1974). Climate Stabilization. *Science* 186:1163–71.

Kendrick, John W. (1961). *Productivity Trends in the United States.* National Bureau of Economic Research, Princeton University Press, Princeton, N.J.

Kennedy, Michael (1976). An Economic Model of the World Oil Market. *Bell Journal of Economics and Management Science* 5 : 540–77.

Koopmans, Tjalling (1973). Some Observations on "Optimal" Economic Growth and Exhaustible Resources. In H. C. Bos, H. Linneman, and P. de Wolff (eds.). *Economic Structure and Development: Essays in Honour of Jan Tinbergen*. North-Holland, Amsterdem, The Netherlands. Pp. 239–55.

———— (1974). Notes on Elasticities. Unpublished memorandum.

Lipsey, R. G., and Kelvin Lancaster (1956). The General Theory of the Second Best. *Review of Economic Studies* 24 : 11–32.

Machta, Lester (1972). The Role of the Oceans and Biosphere in the Carbon Dioxide Cycle. *Nobel Symposium 20* pp. 121–45.

————, and G. Telegados (1974). Climate Forecasting. In W. N. Hess (ed.). *Weather and Climate Modification*. New York.

MacIntyre, Ferren (1970). Why the Sea is Salt. *Scientific American* 223 : 104–15.

McKelvey, V. E. (1960). Relation of Reserves of the Elements to Their Crustal Abundance. *American Journal of Science* 258-A : 234–41.

———— (1972). Mineral Resource Estimates and Public Policy. *American Scientist* 60 : 32–40.

Manabe, S., and T. T. Wetherald (1967). Thermal Equilibrium of the Atmosphere with a Given Distribution of Relative Humidity. *Journal of Atmospheric Sciences* 24 : 241–59.

———— (1975). The Effect of Doubling CO_2 Concentration on the Climate of a General Circulation Model. *Journal of Atmospheric Sciences* 32 : 3–15.

Matthews, W. H., W. W. Kellogg, and G. D. Robinson (1971). *Man's Impact on the Climate*. MIT Press, Cambridge, Mass.

Menard, H. W., and George Sharman (1975). Scientific Uses of Random Drilling Models. *Science* 190 : 337–43.

MRG (1977). Report of the Modeling Resource Group of the National Research Council, Committee on Nuclear and Alternative Energy Systems, Washington, D.C. forthcoming.

National Petroleum Council (1972). *An Initial Appraisal by the Oil Supply Task Group 1971–85*. Washington, D.C.

Newson, R. L. (1975). Response of a General Circulation Model of the Atmosphere to Removal of the Artic Ice Caps. *Nature* 241 : 39–40.

Nordhaus, William D. (1973). The Allocation of Energy Resources, *Brookings Papers on Economic Activity*, Brookings Institution, Washington, D.C. Pp. 529–70.

———— (1976). *Strategies for the Control of Carbon Dioxide*. Cowles Foundation Discussion Paper, Cowles Foundation, Yale University, New Haven.

———— (ed.) (1977). *International Studies of the Demand for Energy*. North-Holland, Amsterdam, The Netherlands.

———— (1978). Computer Code for Bulldog Energy Model, Cowles Foundation, New Haven, Connecticut.

————, and Ludo van der Heyden (1977). *Modeling Technologic Change: Use of Mathematical Programming Models in the Energy Sector*. Cowles

Foundation Discussion Paper, Cowles Foundation, Yale University, New Haven.

Oeschger, H., U. Siegenthaler, U. Schotterer, and A. Gugelmann (1975). A Box Diffusion Model to Study the Carbon Dioxide Exchange in Nature. *Tellus* 27:168.

Pindyck, Robert S. (1977). Gains to Producers from the Cartelization of Exhaustible Resources. *Review of Economics and Statistics*, forthcoming.

PSAC (1965). President's Science Advisory Committee. *Restoring the Quality of Our Environment, Report of the Environmental Pollution Panel.* Government Printing Office, Washington, D.C.

Samuelson, Paul A. (1948). Consumption Theory in Terms of Revealed Preference Theory. *Economica* 15:243–53.

Schneider, S. H. (1976). *The Genesis Strategy.* Plenum Press, New York.

Solow, Robert M. (1974). The Economics of Resources or the Resources of Economics. *American Economic Review*, Papers and Proceedings, 64:1–14.

Stuiver, M., and W. S. Broecker (1975). In WMO (1975).

Taylor, Lester D. (1977). The Demand for Energy: A Survey of Price and Income Elasticities. In Nordhaus (1977). Pp. 1–26.

United Nations (1976). *World Energy Supplies 1950–1974.* United Nations, New York.

WMO (1975). World Meteorological Organization. *Proceedings of the WMO/IAMAP Symposium on Long Term Climatic Fluctuations.* WMO no. 471, Geneva, Switzerland.

INDEX

Cowles Foundation Monographs

16. Harry M. Markowitz, PORTFOLIO SELECTION: Efficient Diversification of Investments
17. Gerald Debreu, THEORY OF VALUE: An Axiomatic Analysis of Economic Equilibrium
18. Alan S. Manne and Harry M. Markowitz, eds., STUDIES IN PROCESS ANALYSIS: Economy-Wide Production Capabilities (out of print)
19. Donald D. Hester and James Tobin, eds., RISK AVERSION AND PORTFOLIO CHOICE (out of print)
20. Donald D. Hester and James Tobin, eds., STUDIES OF PORTFOLIO BEHAVIOR (out of print)
21. Donald D. Hester and James Tobin, eds., FINANCIAL MARKETS AND ECONOMIC ACTIVITY (out of print)
22. Jacob Marschak and Roy Radner, ECONOMIC THEORY OF TEAMS
23. Thomas J. Rothenberg, EFFICIENT ESTIMATION WITH A PRIORI INFORMATION
24. Herbert Scarf, THE COMPUTATION OF ECONOMIC EQUILIBRIA
25. Donald D. Hester and James L. Pierce, BANK MANAGEMENT AND PORTFOLIO BEHAVIOR
26. William D. Nordhaus, THE EFFICIENT USE OF ENERGY RESOURCES

Orders for Monographs 12, 14, 16, 17, 22, 23, 24, and 25 should be sent to Yale University Press, 92A Yale Station, New Haven, Conn. 06520, or 20 Bloomsbury Square, London WC1A 2NP, England.

Orders for Monographs 13 and 15 should be sent to John Wiley & Sons, Inc., 605 Third Avenue, New York, N.Y. 10016.

Orders for Monograph 21 should be sent to University Microfilm, 300 North Zeeb Road, Ann Arbor, Michigan 48106.